Lunar Commerce

Lunar Commerce

Derek Webber

Lunar Commerce

A Primer

 Springer

Derek Webber
Spaceport Associates
Damariscotta, ME, USA

ISBN 978-3-031-53420-1   ISBN 978-3-031-53421-8   (eBook)
https://doi.org/10.1007/978-3-031-53421-8

This Springer imprint is published by the registered company Springer Nature Switzerland AG
The registered company address is: Gewerbestrasse 11, 6330 Cham, Switzerland

Paper in this product is recyclable.

To the Dreamers, Doers, and Risk-Takers of the Lunar Commerce Imperative

—Ad Astra Per Negotium!

Acknowledgments

Let me tell you about retirement. It's a cruel mirage. The work continues, or even increases, and now you don't get paid. For a decade, I have been a member of this world of committees and volunteerism, but of course I have omitted the big plus, which is that you can focus on only the things that matter to you—and at any time simply walk away if it is not working. And that applies also to everyone else working on the endeavor, so it is indeed a mixed blessing. It requires therefore much more flexibility than previous salaried work environments. So, the work gets done solely because of the personal commitments of those involved, and trust becomes a very important, and maybe the most precious, resource. I have indeed been fortunate in that in my case a series of opportunities, and a succession of trusted colleagues, have entered my life, after I formally retired from my consulting business, to help get the (self-selected) ongoing work (all related to commercial development of the Moon) done.

As a child of Apollo, I have always been motivated to follow up on the challenge created by the risks taken half a century ago when the Apollo guys, many of whom I was later able to meet, went to the Moon the first time around. And now we are preparing to return to the Moon, under very different circumstances. So, this book rounds off for me a decade starting around 2012, and dedicated to the work of building private and commercial access to our celestial neighbor.

Since all of the work described in this book has been conducted in my "retirement" years while operating within volunteer organizations, it is particularly important that due recognition is given to all my volunteer colleagues. It is the only "payment" they will ever get! Work within three different volunteer efforts span the decade in question, and I have tried to adequately and fairly capture those efforts at the appropriate points in this narrative. All these folk have, in their precious spare time, done their part in preparing us for a future lunar economy. They clearly believe the Robert Browning dictum: "Ah, but a man's reach should exceed his grasp, or what's a heaven for?"

So, I offer my acknowledgment and thanks for friendship and committed hard volunteer work from the following:

At Google Lunar XPRIZE Judging

Thanks to my fellow international volunteer GLXP Judges: Alan Wells, Jeff Hoffman, Elisabeth Morse, Berengere Houdou, John Zarnecki, Derek Lang, Chuck Reynerson, Jay Kurtz, Dave Swanson, and Craig Peterson. Thanks also to the XPRIZE Foundation professional management and staff: Peter Diamandis, Anousheh Ansari, Andrew Barton, Nathan Wong, David Locke, Chanda Gonzales, and Amanda Stiles. Your combined work gave hope to a whole new generation of potential space engineers and dreamers around the world. And in a very real sense helped open the way to an era of massive cost reductions for lunar activities.

At the ForAllMoonkind NGO

Thanks to the leadership and extraordinary can-do approach to protecting and preserving the lunar heritage artifacts exhibited by Michelle and Tim Hanlon, Marlene Losier, and Judith Beck, my colleagues at our June 2019 presentation to the UNCOPUOS delegates in Vienna. After having succeeded in getting appropriate legislation included in the canon of US law, the work goes on in the difficult international forums where the wordsmithing of treaty language takes place. The lunar tourists are coming, so we need to get these protections in place. We only get one chance to preserve the record and legacy of Earth's first visitors to another world. And thanks to the guys who put the artifacts and boot-prints there in the first place, giving rise to the need to protect and preserve. I thank you collectively for your original risk-taking, and on a personal level to many for your subsequent support to my various lunar commerce related initiatives. It has been an amazing era to be alive, as mankind moved outward for the first time.

At the Moon Village Association NGO

Most of the quantitative, and indeed qualitative, work in this book was conducted by a band of international volunteer analysts, during the period of the global covid pandemic, and so I was never able to meet them in person. We performed our work via Zoom, and I gave my thanks in that same way, but their names need to be recorded and acknowledged here as the creators of the Lunar Commerce Portfolio (LCP). Some were there for the whole 2-year duration of the effort; others contributed as they were able. My apologies if I have missed out any of the contributors. Thanks, therefore, to: Jenna Tiwana, Gidon Gautel, Christophe Bosquillon, Dallas Bienhoff, Sara Sabry, Sylvester Kaczmarek, Vedang Acharya, Erik Kulu, Yann Perot, Enrico Trolese, Andrey Lopantsev, McLee Kerrole, Alistair Schofield, Aditya Schrikhande, Aitor Hernandez, Alexis Caratozzolo, Alvaro Pubill, Antonio Fois, Ehsan Razavizadeh, Norbert Naskov, Ryuichi Dunphy, Sunny Narayanan, Richal Abhang, Varsha Shankar, Vitalii Stoliarchuk, Bryan Lachica, Habeebullah Akorede, Keerthana Gunaretnam, Nidhi Vasaika, Pinar Tan, Samantha Falcucci, Sonalli Madhanraj, Stella Tkatchova, Vishal Tripathi, Pranay Shah, Richard Howard, Jose Ocasio-Christian, Aarohi Khanna, Guillaume Videloup, Laurie Wiggins, Luca Buzelli, Luca Kiewiet, Niesha Baker, Tejas Bendre, Devanshu Jha, Kaori Becerril, Matthias Frenzl, Natasha Heidenrich, Pablo Arellano, Deep Patel, Elaine Tan, Lucien Bildstein, Oscar Fernandez, Veronica Moronese, Ilankuzhali Elavarasan, Keerthana Gunasretnam, Kristine Atienza, Suchwesna Patil, Venkataswamy Eswarachari. In addition to the volunteers, I also want to recognize the administrative and management support of the MVA's leadership board, in particular Giuseppe Reibaldi, Jeff Mankins, Oleg Ventskovsky, and Glafki Antoniou. The work goes on, and new co-chairs have taken over from where I left things, and so I wish them as much luck as was afforded me in having the helpful, creative, hardworking, and trustworthy colleagues, who "got the work done" as reported in this book. Thanks also, for their ongoing LCP support, to the Bocconi team in Milan of Simonetta di Pippo (former Director of the UN's Office of Outer Space Affairs), Andrea Conconi, Mattia Pianorsi, Clelia Iacomina, and Aristea Saputo. And to Mehmet Sefer of Space Construction Technologies, who has taken on the responsibilities of Membership Secretary of the new Lunar Commerce User Group (LCUG).

And Then, There Were The Others

After my having drawn attention to certain conservative elements within NASA with regard to some reticence and difficulties in implementing the task of returning to the Moon—"this time to stay"—I really ought to recognize and give thanks to those who have indeed taken the steps within the Agency to make change happen—to Lori Garver and Phil McCalister, who in their respective roles have led the charge toward the new age using the paradigm-shifting "newspace" approaches. And to that I must add thanks and recognition to the membership over the years of the LEAG and LSIC groups, who have steadily progressed, despite the constraints and vicissitudes of the budget cycle, and tested out many of the basic elements of the potential lunar surface infrastructure referenced in the LCP and in this book.

Thanks to Arlene Kelley for her clear design work and her illustration of the lunar surface tourists.

Acknowledgment is due to the architectural firm of Skidmore, Owings and Merrill (SOM) for their permission to use the image for the cover, which showed the result of a 2020 joint MIT/ESA/SOM collaborative effort at designing a Moon Village.

Thanks to the support team at Springer—Niza Jones and Jill Balzano who accepted my proposal and guided the content, and Harathi Ramu and colleagues who saw to the production process. And in that context, heartfelt thanks to the anonymous reviewers of my original proposal, who provided some excellent suggestions which improved the resulting end product.

Thanks, as ever, to SLF for review and support. I really mean it this time—now I am retired for sure. The boat is waiting, and the lakes of Maine are calling.

August 2023   Derek Webber
                                                                Damariscotta, Maine, USA

Contents

About the Author

Derek Webber Derek Webber is founder of the consulting firm Spaceport Associates.

Originally, as a satellite and launch vehicle engineer, and later in marketing, regulatory, planning, financial, and procurement roles, he made contributions during decades in the satellite communications business sector in Europe, and also later when he came to the United States. He then re-focused his consultancy for the first decade of this century on building the regulatory and business basis of the space tourism industry and through leading a key project directing statistically valid market research among millionaires established the validity of a space tourism market.

Since his retirement, he has been working for a decade on a volunteer basis at various aspects of the commercial development of the Moon, attempting to provide a firm foundation for this important first step of the economic development of the solar system. This has involved initially being Vice Chair of the independent panel of Judges for the Google Lunar XPRIZE, an international competition to encourage non-governmental private low-cost access to the Moon, and then in managing volunteer team members in two NGO organizations which have Permanent Observer status at the Vienna-based United Nations Committee on the Peaceful Uses of Outer Space (UNCOPUOS). The first of these was *For All Moonkind*, where the aim was the preservation of lunar heritage sites, during an expected upcoming era of renewed surface activity on the Moon, including sightseeing by lunar tourists. The second NGO was *The Moon Village Association*, where during a two-year period he formed, and served as co-chair of, the Working Group on Lunar Commerce and Economics. Within this group, he directed team members performing the analysis involved in creating the *Lunar Commerce Portfolio (LCP)*—a first attempt at a fully transparent assemblage of data on lunar market segments, and deriving resulting demand-based lunar revenue forecasts. This book captures the results of this last decade of lunar commerce related activities, and includes within it an easy introduction to the purpose, procedures, and content of the LCP.

Derek Webber is the author of four previous books on commercial space development and space history, describing the satellites, launch vehicles, the Moon landings, and the space tourism sectors.

Acronyms

A

A&E	Architecture and Engineering
ASA	Australian Space Agency

B

BLSS	Bio-regenerative Life Support System

C

CAB	Chargeable Atomic Batteries
CLPS	Commercial Lunar Payload Services
CNES	French National Centre for Space Studies
CNSA	China National Space Agency
COPUOS	Committee On the Peaceful Uses of Outer Space
CSA	Canadian Space Agency

D

DARPA	Defense Advanced Research Projects Agency (of US Govt)
DLR	German Space Agency
DOE	Department of Energy (of US Govt)
DSN	Deep Space Network (of NASA)

E

ELDO	European Launcher Development Organisation
ESA	European Space Agency
ESM	European Service Module (of Orion spacecraft)

ESRO	European Space Research Organisation
EUMETSAT	European Meteorological Satellite Organisation
EUTELSAT	European Telecommunications Satellite Organization
EVA	Extra Vehicular Activity

F

FAA	Federal Aviation Administration
FCC	Federal Communications Commission

G

GEGSLA	Global Expert Group on Sustainable Lunar Activities (of MVA)
GLXP	Google Lunar XPRIZE
Govt	Government
GPS	Global Positioning System
GROWLER	Grading and Rotating for Water Located in Excavated Regolith

H

He3	Helium 3
HiDef	High Definition
Hi-Vac	High Vacuum
HLS	Human Landing System (of Artemis program)

I

ILRS	International Lunar Research Station
INMARSAT	International Maritime Satellite Organization
IOAG	Interagency Operations Advisory Group
ISA	Israel Space Agency
ISRO	Indian Space Research Organization
ISRU	In Situ Resource Utilization
ISS	International Space Station
ITA	International Trade Administration
ITU	International Telecommunications Union
IWP	Ionomer Membrane Water Processing

J

JAXA	Japan Aerospace Exploration Agency
JPL	Jet Propulsion Laboratory (of NASA)

K

KREEP   Potassium, Rare-Earths, Phosphorous

L

LCE   Lunar Commerce and Economics (Working Group of MVA)
LCP   Lunar Commerce Portfolio (of MVA)
LCROSS   Lunar Crater Observation and Sensing Satellite
LCUG   Lunar Commerce User Group
LEAG   Lunar Exploration Analysis Group
LEM   Lunar Exploration Module (of Apollo program)
LMV   Lunar Mobility Vehicle
Low-g   Low Gravity
LRO   Lunar Reconnaissance Orbiter
LRV   Lunar Roving Vehicle (of Apollo program)
LSA   Luxembourg Space Agency
LSIC   Lunar Surface Innovation Consortium
LTV   Lunar Terrain Vehicle

M

MAPP   Mobile Autonomous Prospecting Platform
MoU   Memorandum of Understanding
MVA   Moon Village Association

N

NASA   National Aeronautics and Space Administration
NGO   Non-Government Organization
N/K   Not Known
NZSA   New Zealand Space Agency

O

OST   Outer Space Treaty

P

PGM   Platinum Group Metals (Ruthenium, Rhodium, Palladium, Osmium, Iridium, Platinum)
PhilSA   Philippine Space Agency
PPS   Peaks of Perpetual Sunlight
PSR   Permanently Shadowed Region

R

RDI	Required Daily Intake
RF	Radio Frequency
RGD	Radiant Gas Dynamic
ROM	Rough Order of Magnitude
ROSKOSMOS	Russian State Space Corporation
RTG	Radioisotope Thermoelectric Generator

S

SEE Lab	Space Economy Evolution Laboratory (of Bocconi University)
SEI	Space Exploration Initiative
SLS	Space Launch System
SMR	Small Modular Reactor

T

TBD	To Be Determined

U

UAESA	United Arab Emirates Space Agency
UNCOPUOS	United Nations Committee on the Peaceful Uses of Outer Space
USG	United States Government

V

VSE	Vision for Space Exploration

W

WG	Working Group

X

XPRIZE	Xprize Foundation

1

Introduction

Been there, done that? Why go back to the Moon—didn't we already do that 50 years ago with Apollo? We didn't find anything worthwhile up there, or we would have stayed, right? What on Earth, if you'll pardon the expression, will folks be doing there this time around to make it worth the effort, and to be such an expensive distraction from other more urgent matters? What has changed?

The general public (that means all of us) can be forgiven for raising such questions, because there has been a rather poor—and even confusing—narrative associated with the current US-led Artemis Program activities. On the one hand, there is talk about this time we are going to the Moon to stay, and that "sustainability" is therefore a key aspect of the planning. However, the funding being provided to the governmental space agencies to make this happen is only sufficient for a few trips to the Moon, even years apart (whereas during Apollo, when the intention was not even to create a permanent presence, the crews nevertheless went twice a year throughout a four-year period). Also, back in the sixties, there was complete clarity among the public (and of course the astronauts) about why we were going. It was about science and exploration, but mainly about the USA beating the Soviets in an existential battle for dominance in the new medium of space. Nowadays, the public does not in general consider (do we?) that there is a geo-political rationale for going back to the Moon. Certainly, the Chinese and other nations are continuing to send spacecraft, and human crews will eventually be included, but surely this cannot be a race, if the USA already won the race a half century ago? And *commerce*? Really? What are we even talking about?

In this book, we are going to come to terms up-front with these questions and provide the missing rationale. We'll explain what's new, and how recent data on lunar resources has opened up expanding possibilities. We are,

D. Webber, *Lunar Commerce*, https://doi.org/10.1007/978-3-031-53421-8_1

furthermore, going to make it clear how this new generation of lunar activities will benefit all of us on Earth. Spoiler alert—we go to the Moon to make life on Earth better, and in the long term, ultimately even possible.

And once the rationale is clear, this will lead us inevitably to the idea of a joint governmental and commercial approach, which, in turn, leads us to investigate how we might attempt to create a viable lunar economy. *We are talking about the birth of lunar commerce.* You might ask how can anyone possibly intend to do business on the Moon. Who will be selling what to whom? Come on, can we really expect to build a marketplace on the Moon? I have answers for you. The bottom line is "yes." This book will help us understand the likely scale and timing of business activities on the Moon. It will scope out which lunar markets will likely be part of the solution, and to what degree they will be truly commercial, and not dependent on government funding (or, in other words, on our taxes). This is a first step, maybe a bit bold, maybe even a bit halting. But a start.

Some have strongly held arguments against the very idea of pursuing this approach of the commercial development of the Moon. We'll talk about that, as well. We have a modest goal here—simply to explore the possibilities and then provide the best current assessment of potential lunar business revenue opportunities, using a rigorous and fully documented approach, and thereby providing a vocabulary and framework for future integrated commercial lunar endeavors. That may sound like a mouthful, but there is really no alternative if we want to have a firm foundation for building a lunar economy. There's still a long way to go. An important aspect of this work is the need for a realistic assessment of uncertainties. You'd better believe it, there's a whole lot of guesswork involved, and some really flaky data. But, as Francis Bacon said, a long time ago, "If a man will begin with certainties, he shall end in doubts; but if he will be content to begin with doubts, he shall end in certainties."

The whole point is to make it possible to improve, and get a more confident idea of what the future holds—of what the potential is for business on the Moon. We'll make it clear where the data and key assumptions are lacking. We'll point out where there is a need for market research and more testing of technologies to reduce the uncertainties and therefore the associated risks. Back in the day, with the Apollo astronauts, it was all about *going*—faster, higher, and farther. Now it's declared to be about *staying* there—and what that will take.

We are only at the very beginning of efforts at commercializing the Moon. There are enormous gaps in our knowledge. How can we possibly know about lunar business potential decades into the future? Commercial operators and their funders at present (maybe you are thinking of coming on board) nevertheless have to understand something of the scale of the risks they are undertaking, and importantly how to proceed to make the risks more manageable. So, this book can only be a primer. It is a primer in two senses. First, it introduces the very idea of creating a lunar economy, and what that might entail, and how that is different from simply having the government pay for everything. Then, it more specifically provides the background, and thinking, behind the *Lunar Commerce Portfolio (LCP)*—a document and model created within the Moon Village Association (an NGO with permanent Observer status at the UN's Committee on the Peaceful Uses of Outer Space, based in Vienna) for future users of the lunar commerce domain. It contains the first integrated revenue estimates, together with their associated transparent assumptions and enormous range of uncertainties. Think of it as a reality check. Maybe it's a bit too soon to break into the piggy bank? But, perhaps its most important contribution is to pull together a skeletal framework of integrated market activities, populated at present with only our first assessments of values. It is hopefully a platform to allow us to begin to build a rigorous database of lunar commerce opportunities. The framework is intended to be overarching, and to be capable of a succession of future data updates, revisions, and improvements. It's a basic reference structure for future contributions toward a gradual improvement in understanding lunar commerce. The assessments of lunar commercial revenue opportunities can only get better as newer data becomes available. We've never tried to live on another celestial body before. So, we have no prior reference guide. We are going way beyond Neil's "one small step" in this grand new initiative. This work represents a first stab at coming up with the supporting business perspective.

There are a number of different groups who will be awaiting these lunar business revenue assessments. Certainly, the potential commercial operators themselves need this information. And so do their funding sources and insurers. The governmental space agencies around the world need the information, too. Why? Because they are assuming that their governmental budgets will be *augmented* by commercial revenue streams, but have very little understanding of what the likely scale, or timing, of such streams might be. You will discover how important to the outcome of lunar commerce development are the potential actions of international and national regulators. So, these regulating agencies also need to have access to the best available revenue data, and in particular need to know how their actions or inactions in various policy areas will impact the scale and timing of the lunar commerce revenue streams.

Although this book is a primer, and therefore can't be expected to contain *all* the individual detailed information, values, and sources involved in deriving this first-cut at lunar commerce revenues, it is based on the aforementioned major work which does indeed contain this full level of detail. This book, therefore, while describing other things that transpired in the decade covered, is also designed to provide an easy introduction to the content of the Lunar Commerce Portfolio, created by an amazing motley crew of international volunteers in a Working Group of the Moon Village Association. It also incidentally includes *a first assessment of lunar space tourism demand.* The first edition of the LCP (Moon Village Association, 2022) was issued in November 2022 and presented in Los Angeles, at a joint forum of the National Space Society (NSS) and the Moon Village Association, by the author who was a co-chair of the MVA Working Group on Lunar Commerce and Economics which created the work (ably assisted in person at that event by Working Group colleague Dallas Bienhoff, and supported virtually by a bunch of other analysts around the world by Zoom call). The work of improving the content of the LCP continues within the Moon Village Association and elsewhere, and will eventually be represented by the issue of second and subsequent editions, but in this primer, we refer only to the First Edition (and therefore reflect 2022 assumptions, data, and values). Keep your eyes open for the updates. But this primer will continue to be the foundational initial hardcopy reference source for understanding the rationale for the LCP.

There are two cut-off dates pertinent to the decade's worth of material contained in this book. First of all, the content of the Lunar Commerce Portfolio Version 1 was publicly released in *November 2022*, and so all references within this book to that specific material do not reflect any subsequent events. In preparing the other background material in the remainder of the book, because of the rapid changes taking place in this commercial workspace, a later cut-off date of *August 2023* has been used.

And just to provide all of us with a sense of perspective, Fig. 1.1 shows what we are attempting to achieve (thanks to an image provided by the Chinese spacecraft Chang'e 5 from beyond the Moon in 2020). The Earth in the background is 400,000 km (i.e., 3 days' journey) from the Moon. Twenty-four NASA astronauts made that journey back in the sixties and early seventies, risking their lives. They made it possible for us to even have the discussion contained in this book. Those of us who have come later in effect are continuing their endeavors. The motivation may have changed to some degree, as a new generation of Moon explorers are getting ready for their Artemis missions. But there can be no doubting of their courage as they set forth anew. Let's see how, this time, we can make it truly relevant for all of us. We'll cover the Why and How of returning to the Moon, and how this time we are going to turn things around so that the Moon makes money for us, potentially for all of us, if we can get behind the effort.

Fig. 1.1 The vast distance from the Earth to the Moon, where we are attempting to build a self-sustaining economy. (Credit: www.news.cn China National Space Administration)

We have done our best to create a valid and analytically sound portfolio of opportunities and revenue estimates for a self-sustainable lunar economy. The ground work is being put in place. So, let's go!

Reference

Moon   Village   Association.   (2022).   Moonvillageassociation.org/download/ the-lunar-commerce-portfolio-first-edition-november-2022

Part I

Why We Go

This section seeks an explanation of why we are even considering lunar commerce as a possibility.

It emerges that there are both (very) long-term and much shorter-term rationales. In both cases, the beneficiaries would be here on Earth, and in the shorter-term there are real possibilities for business growth.

The Lunar Commerce Portfolio is introduced as a framework for understanding the potential lunar markets, their revenue opportunities, and timeframes.

There is nevertheless an acceptance that some objections will need to be addressed before the revenues can be realized.

"Teach thy necessity to reason thus:
There is no virtue like necessity"
– Shakespeare, King Richard II

Part I

2

A New Industrial Revolution

So, let's consider briefly what are the political, social, economic, military, and even existential reasons for going back to the Moon, and why we are even considering undertaking its commercial development. We could start with why we even send robot spacecraft into space at all, and then think about why humans need to go there. Then we can try to understand the particular role of the Moon in the future of mankind's development, and even existence.

I come from a shipbuilding community in the North-East of England, and now live in a town in the state of Maine in the USA where wooden schooners were once built. I also grew up immersed in the British schoolboy backdrop, setting high values on heroic risk-taking explorers such as Captain Cook and Shackleton. So, I know about "going forth." We go because we can. Whether on land, sea, or in the air and beyond. As Dave Scott, the commander of Apollo 15, proclaimed as he descended the ladder onto the lunar surface: "Man must explore." So that's one reason right there.

As a natural evolution of the development of flying and rocketry, which started simultaneously for all intents and purposes in 1903, we began seriously going into space in 1957 with Sputnik 1, basically because it was viewed as a new domain for human knowledge, scientific knowledge. Folks at the time talked about the conquest of space. There was a strong military underpinning, too, leading to the "space race" of the sixties, where the Soviet Union and the USA were competing to understand and develop the technologies needed to operate in this new domain. The origins, after all, were military. The rockets' red glare was a part of battles, even before the V2 missile emerged during the Second World War. After the first sounding rockets and satellites, Gagarin and his successor astronauts followed into space. Why did they think they were doing it? For sure, there was the competition between military test pilots on both sides of the divide in the Cold War.

But there were loftier ideals, too, some even going back as far as the early Russian rocket theorist Tsiolkovsky at the end of the nineteenth century (Tsiolkovsky, 1903). From the perspective and vantage point of Earth orbit, the notion emerged that the Earth has problems, even though it remains, so far as we know, by far the best place for humans (and all life) to be. Some of the potential problems are very long term; others might be more immediate. Among potential long-term natural calamities that might befall the Earth are large asteroid impact, gamma ray bursts, magnetic pole reversal, super volcanoes, and changes to the sun. Neil Armstrong pointed out that the Earth's environment is not benign as our planet continues in its annual procession around the Sun and: "I have had the privilege of looking down from above the atmosphere and have seen shooting stars far below me" (May, 2014). And we can add to that list the possible man-made risks of nuclear war, biological agents, increasing population, diminishing raw material stocks, global warming, and pollution.

The good news is that our space efforts can provide solutions. Our half century of space exploration has shown us the potential problems, but also the route to solutions. If an asteroid is a threat, can't it also be viewed as an opportunity? We can maybe redirect it to remove the threat, and mine it for minerals to help solve a terrestrial resource problem. That would be neat! Space, and particularly our solar system, can be a future source of resources and energy for mankind's future. And in the extremely long timeframe, it can provide us with a "back-up plan" for Earth, in response to the existential cataclysmic events we listed. That's certainly the way Buzz Aldrin sees it. And other Apollo guys too, like Charlie Duke and Ed Mitchell, who have both shared with me their strong support for getting back to the Moon, and specifically using space tourism as a means of enabling it. Which is where the notion of "settlement" comes in. And the need for a "space-faring civilization." We made a start in 1969 when Neil and Buzz set foot on the Moon's surface, even though it looked unpromising. Didn't Buzz call it "Magnificent Desolation" (Aldrin, 2009)?

After the Apollo landings, we rapidly learned how to send robotic exploration missions throughout the solar system, and knowledge of astronomy in particular was transformed by these endeavors. Remember Hubble? Remember that in the last half-century we sent spacecraft with cameras and instruments to visit *all* the planets and moons way out to Pluto? And, we also learned a great deal about the Earth, its atmosphere and its oceans, and unfortunately about how humans have been polluting and endangering our home planet.

Then there was the commerce. Apart from the new employment opportunities that emerged at spaceports and rocket and spacecraft manufacturers, entire new industries emerged, such as satellite telecommunications, broadcasting, and navigation. We began to make real money out of space. This is how today your gizmo can direct you to the nearest McDonalds. Space became a new area of economic development, not always dependent on governmental tax-funded operations. To the tune of several hundred billion dollars a year.

We can't of course expect terrestrial political and budget systems to address the extremely long time periods involved (maybe hundreds of thousands, or millions, of years for some of these events), even when ultimately human survival (and maybe even the survival of all life) is at stake. But it does need to be understood in the background, as the ultimate rationale for our space activities, and particularly those involving human crews. It is truly why we go. And it does need to color our general incremental sense of direction in what we are doing. But we do realize that we'll need much nearer-term benefits to justify these ongoing space development activities to the general public, in the context of normal political and economic discourse here on Earth. We shall need better, and more immediate, answers to the awkward questions you raised at the start of this book. We have not fully answered them yet.

And this is where the Moon comes back into the picture. Our celestial neighbor is too close to the Earth to be the solution and response for *all* of the potential astronomical calamities we have listed—it will for example eventually suffer the same fate as the Earth when the sun expands during a later stage of its life cycle. We would need to go further out, away from the Sun. So, No to Mercury and Venus, Yay toward Mars! The Moon is however an ideal first location—only three days away—to try out humanity's long-term imperative of the greater settlement of our solar system (setting the stage for future economic activities on Mars, asteroids, and the other moons of the planets). The work of Apollo is unfinished. The Moon is in scale the size of a new continent on Earth, and we have only begun to explore its potential, and the new knowledge that it contains. And in particular, the Moon presents us with the opportunity to experiment with what would be involved in permanent human settlement at much greater distances, *and the commercial activities that could support such an enterprise.* We need to take onboard the idea of in-situ resource utilization (ISRU). Making something useful out of Moondust. This is important, but it still does not answer the main question from the man/woman-in-the-street, about what's changed, and what's in it for me? And what is it that people mean when they talk about "the lunar economy"? Can we repeat the trick of the first half century of mankind's engagement with space, when so much business was created, and so much creative energy was profitably employed by the "traditional" commercial space businesses? Being able to watch Formula 1 motor racing, the Red Sox game, and the Tour de France live all over the world is so cool. What can possibly be a follow-up act?

How about this. I grew up in Northern England, where the industrial revolution began—steam engines, locomotives, which went on to change the world in an earlier century. Maybe we can think of this current opportunity as a potential *new industrial revolution* opening up. Going back to the Moon to stay could be the start of a massive change. There will be new players. You could ride this wave. This is an open forum for developing new technologies and creating a continent-sized zone of development (depending, of course, on what emerges from the regulatory agencies). What's changed is that, among other things, new findings about lunar resources can now label the Moon as an orbiting gas station of the Earth (it contains the raw materials of rocket fuel) for future human and robotic space travelers. It conveniently sits in a relatively small gravity-well dimple just three-days' journey from Earth. It contains a continent-sized source of minerals and oxygen within its rocks, and a so-far unexplored source of water in newly-identified cold traps (permanently shadowed regions, or PSR's) in its polar regions which could form the basis for creating that rocket fuel—and of course for water and oxygen for lunar residents. What's changed is that new techniques—reusability of rockets, in part conceived to make space tourism possible, private home-built spacecraft (as encouraged by the Google Lunar XPRIZE competition), the "cubesat" revolution, etc.—have made the costs of doing business in space much more manageable. So, this begins to answer the basic question. We can maybe use the Moon to make money, rather than see it as a drain on resources.

Now that's a compelling reason for our interest, support, and involvement. It becomes a particularly strong argument, when we also include the aspect of access to scarce (and possibly militarily important) resources. It is definitely a nearer-term proposition than the "back-up plan for Earth" rationale. But we need to be honest and open with all of us taxpayers about a timeframe which could still extend a few decades before producing any positive returns independent of direct payments from the government. Some countries, such as China, understand this long-term view. Of course, as was mentioned earlier, not everyone thinks this should in any case be done, for a variety of cultural and social reasons (to be discussed later). But we humans have done this before. We have found ways to make money out of space. Satellite broadcasting and communications have made big bucks for decades (certainly helping put my own daughter through school). Space tourism is starting to make money too.

Lunar commerce could be the next killer app. This is a new movement; a new moment. Countries are rushing to get on board by building spaceports.

Universities are creating courses. Have you, for instance, heard what Luxembourg is doing? They plan to be the world's leading country when it comes to mining asteroids. So, in short, we go to the Moon to help make life better down here. Economic growth. New possibilities. Somewhere to find resources that will save us from further environmental damage here on Earth. A place to train for the ultimate need for settlement off-Earth. But don't fret about that last one during the next budget cycle.

Who can lead in this new phase of mankind's venture into space? How can we address these very long-term issues within the short-term attention span of folks, and budget cycles, today? At least two of today's US titans of industry (Elon Musk of Tesla and SpaceX, and Jeff Bezos of Amazon and Blue Origin) fortunately grew up being motivated by the achievements of Apollo, and have dedicated a great deal of their time and money to becoming a part of the next steps, the commercial steps. Interestingly, they are each targeting a different aspect of the problems of Earth, and of using space as a source of solutions. We need them both to succeed. Musk is working on the "backup plan for Earth" part of the problem, and has built his Starship (Starship, 2023) with the intention of using it to take possible settlers to Mars (Musk, 2015). Bezos, on the other hand, is aiming to try to solve Earth's resource and pollution problems, by taking heavy industry into space, so that Earth can remain cleaner and habitable for much longer (Bezos, 2016). Both of them, nevertheless, have made parallel first steps—making access to space reusable, and therefore very much cheaper and more reliable, for all future uses. Also, significantly for our purposes, they both recognize the very long time-scales involved, but see the commercialization of the Moon as the next phase, and are heavily invested in returning to the Moon by supporting the US-led Artemis program, and providing commercial lander services to the governmental space agencies, and commercial players, wanting to go. If indeed we have messed up the Earth and don't want to repeat the process on the Moon, as some would claim, we need to figure out how to this time do it safely, carefully, cleanly, peacefully, and maybe even more equitably. These are global questions needing global answers. Every time I walk past the UN headquarters in New York City, I am reminded of this new great challenge we all must face together. We can clean up the place (i.e., Earth), make a new great economic engine, and guarantee our long term future, all at the same time. OK, it might take a while. But it is surely worth the try for such prizes.

So, lunar commerce, here we come! Fortunately, some progress has been made into understanding what is required to work on the Moon, and in particular, some demonstrations of In-Situ Resource Utilization development have begun to take place (ISRU, 2023). This has been happening using government funding, as appropriate in the early stages of a massive development endeavor. But at some stage, the true commercial investment will need to take place. Commercial operators and funding sources have a different set of guidelines than do governments, and they will quite naturally keep waiting for more and better data from the government and academic experimenters in order to minimize their risk before committing significant resources. However, if there is a potential great opportunity for commercial success, maybe in scale far bigger than for other possible opportunities on Earth, then the new lunar operators will commit to becoming part of the lunar economy, and will engage in the normal commercial competitive activities. You could buy a piece of the action yourself. Did you miss out on the dot.com boom investments, perhaps? In principle, your investments will contribute to gradual reductions in prices and increased efficiencies within the new lunar market sectors. So, we see again why the quantification task described in this book (and especially its assessment of risk and likely time scales) is important.

A great deal of steady background research and progress has been achieved since we were last on the Moon, and indeed, it continues thru working groups and committees, often supported by NASA funding, via organizations such as LSIC (Lunar Surface Innovation Consortium) and LEAG (Lunar Exploration Analysis Group) (LSIC, 2023; LEAG, 2023). You probably didn't even know it was going on. It wasn't a secret, much less a conspiracy, it just did not seem to fit into any plan, or at least any published public domain plan, that made practical sense. I stumbled across these efforts once I had decided to try and build the case for a lunar economy, and I can report that they are welcoming of new members. Now we know that they exist, and indeed why the whole concept is so important—and are glad of this seemingly stealthy progress which has been taking place. Just so you know, LEAG has been operating since 2005, or earlier, and is supported thru NASA, while LSIC is more recent, starting in 2020, and functioning thru Johns Hopkins University. Even more recently, a new center for studying the needs and results of lunar resource excavation (in fact they focus also on asteroids) has been established in 2021 at European Space Resource Innovation Center (ESRIC) based in Luxembourg, from where they host an annual "Space Resources Week" discussing the latest thinking and findings. LSIC, for example, has a membership consisting of 26% government, 41% industry, 22% academia, and 12% nonprofit, and operates via a series of focus groups, including working groups on lunar dust mitigation, surface power, In-Situ Resource Utilization, and Excavation and Construction. Examples of topics studied within the LSIC forum over its first two years of operation included lunar simulants, power systems, dust mitigation, terrestrial analogs, ISRU testing, resource potential modeling, standardization and interoperability, robotic mining, lunar mining production processes, sintering (i.e., forming a solid surface by heating and compression) of regolith, habitat construction and outfitting, and discussions of the upcoming NASA CLPS missions and experiments (discussed later). And plenty more, too.

Will there emerge ways to balance the needs to protect the lunar environment, with the needs to use the Moon as part of humanity's very long-term plan to include the solar system as part of our economic resources? Will the answer to these questions emerge through careful study in academic institutions and in meditative international forums such as at the UN's Committee on the Peaceful Uses of Outer Space (UNCOPUOS)? Or (maybe more likely), will they become a side effect of political and military activities in an area viewed as having some strategic importance? The Moon is a potential answer to shortage on Earth of various critical resources, such as rare-earth metals, and as such may be a potential zone of conflict over the decades as terrestrial resources diminish. We'll need to try to protect against that happening. We do now at least know (through analysis of Moon rocks with newer techniques) that these resources do indeed exist on the Moon. Whatever the answer to these important questions, we need to know what it would take to create a commercially viable human settlement on the Moon, the task that we undertook within the Lunar Commerce Portfolio, to be described later. Someone had to make a start.

You will find (Spoiler alert!) that two things will emerge to be critically important to this endeavor. First of all, a realization that all the potential lunar markets need to develop together *in parallel*, since they turn out to be dependent on each other for success. That will be tough to make happen. Then, it becomes apparent that government needs to *continue to be involved* in the areas of commercial lunar developments, both as the initial funders of high-risk development activities, and as initial customers for the commercial providers. So, don't expect that the space agencies will submit a reduction in their respective budget requests yet, although the subsequent growth needs will, it is hoped, eventually be picked up through commerce.

So, what is this outfit with the weird name? The Moon Village Association, of volunteers, was established in 2017 as an NGO to understand and assist the international aspects of cooperating in the sustainable return to the Moon (Moon Village Association, 2023). The MVA explores legal, regulatory, cultural, economics, and architecture-related aspects of returning to the Moon. The ongoing work of the Lunar Commerce Portfolio, within the Moon Village Association, attempts to determine when the lunar economy can become free-standing without the need for ongoing government subsidy

(Acharya et al., 2022). Feel free to join up and help the ongoing task—simply check out their website. And, of course, all of this will emerge against the background of *realpolitik*. Very high potential rewards come inevitably with very high risks. We have been doing our best to create a way to gradually reduce the uncertainties and associated risks, so that the various decision makers can move forward representing their respective diverse interests. Welcome to the new industrial revolution.

References

Acharya, V., et al. (2022, September). *Lunar Commerce Portfolio: The structure, actors, and revenue potential of the emerging lunar economy.* In IAC-22-D3.1.5x70456, 73rd International Astronautical Congress, Paris.

Aldrin, B. (2009). *Magnificent desolation.* Harmony Books.

Bezos, J. (2016). *Jeff Bezos lifts curtain on blue origin rocket factory, lays out grand plan for space travel that spans hundreds of years.* Geekwire.com (Alan Boyle).

ISRU. (2023). *In situ resource utilization.* https://www.nasa.gov/ISRU

LEAG. (2023). Lunar Exploration Analysis Group. LPI.USRA.EDU/LEWAG/

LSIC. (2023). Lunar Surface Innovation Consortium. LSIC.JHUAPL.EDU

May, B. (2014). *Starmus—50 years of man in space.* Canopus Publishing.

Moon Village Association. (2023). https://moonvillageassociation.org

Musk, E. (2015). Futurism.com/Elon-Musk-we-must-leave-earth-for-one-critical-reason

Starship. (2023). https://www.spacex.com/vehicles/starship

Tsiolkovsky, K. E. (1903). *Investigation of world spaces by reactive vehicles. Selected works.* Mir Publishers, 1968.

3

Making It Happen

We said that we need to return to the Moon for a combination of reasons, some with an exceedingly long timeframe, and others with a more immediate focus (such as those reasons related to using the Moon's resources to replace scarce reserves of certain materials on Earth). Messrs Musk and Bezos are doing their bit on their respective schedules, to address these matters, while making a buck in the process. How urgent is this—is there some kind of window of opportunity in which we need to conduct this work, or do we have multiple decades to bring it to fruition? Opinions on this differ, and we shall merely present the data as the analysis unfolds. The two extreme viewpoints on this may be represented on the one hand, by those who say there is no urgency (and perhaps citing that no inbound asteroids capable of devastating the Earth have been detected that would arrive within the next hundred years, say). Or on the other hand, there are those who view the situation as a window of opportunity, in that we collectively still know how Apollo was done, and furthermore, we are fortunate to have available private capital engaged in space exploration due to the company founders' interest and engagement following those first Moon landings in the 1969–1972 period. And, furthermore, they might importantly add that there may be disasters that befall the Earth more rapidly than we have assumed (e.g., as the result of global warming, or a global pandemic, or magnetic pole reversal, etc.). Since the timeframes for a future disaster scenario vary between on the one hand many thousands, or hundreds of thousands, of years, and on the short-term assumption maybe only a decade or so, then it is best, they would argue, to err on the side of caution and operate as though the issue truly is urgent, and that we therefore need to rise to the opportunity offered by the window—especially since we have demonstrated 50 years ago that we do know how to do it. Make

D. Webber, *Lunar Commerce*, https://doi.org/10.1007/978-3-031-53421-8_3

sense? Do you agree with that thought? Elon Musk, the President of SpaceX, does. In fact, it was his own formulation (Musk, 2015).

So, we are operating with uncertainty about the long-term quality and sustainability of life on Earth due to the depletion of resources that has been caused by humans, global warming, and astronomical and other factors. Because of the relative success of humans as a species, this has in fact always been the case; it is just that we now know about it, and the invention of the rocket has given us an option. There is a potential solution, but it would involve the use of the Moon. We have been to the Moon before, so we know how to do it. However, the traditional Apollo way was just too expensive for the kind of sustainable activities now envisaged (see next chapter). And this brings us back to the need to find some other mechanism and motivation in order to bring it about. Commerce can be the answer. Lunar commerce—made possible because the Clementine findings in 1994 (Clementine, 1994), and the LCROSS experiment in 2009 (LCROSS, 2009), identified the previously unknown water reserves—provides the motivation, and possibly the funds, depending on the findings of the analyses we're going to be reporting in this book. Let's find out. We're going to attempt to create a realistic assessment of the relevant overall scheduling for the commercializing of the Moon. In conducting the analysis recorded in this book, we are offering our contribution toward making it happen.

With this premise, we now have answers to the question posed at the outset about who will benefit and why should you care. In a capitalist society, potentially everyone can benefit from the new business opportunities, and all of humankind eventually benefits by having taken the first steps toward settlement through economic development of the solar system, which could eventually ensure the continuation of human (and indeed all other) life, including of course your own descendants—surely something worth caring about. We have to wonder why NASA and the other space agencies don't simply just say it.

While the reasons are existential in the long term, they affect the quality of life in the shorter term, and we therefore need to figure out how to make this happen. And the work presented in this book kick-starts the crucial commercial support system. It begins the process of determining what are the potential investment opportunities. Furthermore, we'll bring together the data that will demonstrate the opportunities, and assess the factors which would help, and those which would hinder, the establishment of a lunar economy. This material will demonstrate the driving economic reasons that will ignite both active investment and public support. We shall also be introducing our best assessment of who are, or will be, the major actors. And importantly, we stress that this is just a beginning. Spoiler alert—it might be just too tough yet to be able to confidently identify, to the satisfaction of an investor community, where to focus funding resources. But we are going to try.

The findings of this first venture into understanding lunar economics will of necessity be clouded by uncertainties, which in turn are a consequence of the considerable unknowns at present in the basic assumptions about markets and technologies. Future iterations will improve this situation, to everyone's benefit. Perhaps, our greatest contribution here will be to point out just where these unknowns are affecting the outcome, and clouding the ability of investors to decide to proceed. Just knowing what we don't know can be a big plus. And another significant finding will be to place realistic timeframes on making lunar commerce a reality. It serves no useful purpose to fuel the machinery of development with statements, data, and assumptions which are dictated by hype. This can only lead to disappointment and financial ruin. We are going to bring together our best estimate of realistic timeframes in order to make our contribution to making this future happen. Or at least to enable subsequent commercial entities to study the pros/cons of the potential investments using the best information possible within the public domain.

After the best part of a decade thinking about it, I've been doing work with a group of friends within the membership of the Moon Village Association to see what we could find out if we crunched some numbers. They were smart people, but none of them smart enough to foresee the future, so we had to make assumptions. An assumption is a guess, hopefully well-informed. My colleagues had experience from a number of disciplines and came from a wide range of countries, and we used a consensus approach to hopefully arrive at reasonable and wise choices for each necessary assumption. But in any case, we are going to make our assumptions clear, so that they can be challenged, as indeed, they should be.

As we proceed with our task, therefore, we'll be studying *all the potential lunar markets*, and referencing *the suppliers* currently contemplating entering the lunar commerce field, and those *customers* whom they would be likely to find for their product and/or service. You will therefore have at your disposal an effective sourcebook of data to help you in your future investment decisions, including information on the potential competitors also contemplating this market area. You will find *supply chain* information which will suggest places where you might jump on board, either as an investor or as an entrepreneurial product or service provider. In this new lunar economy, it will not be necessary at all stages of the value chain (that's just a supply chain showing where you can add value—Porter, 1985) to be a rocket scientist to succeed. Someone will need to be the barman at a lunar space tourism hotel. At a national level, there will be opportunities for countries having no previous involvement in space activities to opt to be included, and support their citizenry in some chosen part of the supply chain where they have something to offer.

Making it happen requires technological innovation, public support, lots of risk, an agreed international commitment to addressing the global problems, and a viable lunar economy. This work is our contribution to getting the last part going. Do you want to buy into one of these companies? We want to see if this can indeed be a grand new space business opportunity, a follow-on to the satcoms business, and if so to help make lunar commerce happen, within the regulatory bounds of possibility. We warn of course that at this stage there is considerable uncertainty. Furthermore, we only study revenues, and not costs. This is difficult enough. So, we can't advise on profitability. However, it gets us on our way. We now know how we can make it happen.

References

Clementine. (1994). Solarsystem.NASA.gov/missions/clementine/in-depth/

LCROSS. (2009). *LCROSS impact data indicates water on moon*. NASA.gov/mission-pages/LCROSS/main/prelim_water_results.HTML

Musk, E. (2015). Futurism.com/ELON-MUSK-WE-MUST-LEAVE-EARTH-FOR-ONE-CRITICAL-REASON

Porter, M. (1985). *Competitive advantage: Creating and sustaining superior performance*. Free Press.

Part II

How We Go

Having decided that there is indeed an imperative for mankind to return to the Moon for the reasons discussed in Section I, this section compares and contrasts two possible approaches for attempting to do this sustainably.

On the one hand there is the more traditional government-led endeavor, as exemplified by Project Apollo historically, and the current Project Artemis missions.

And, then there is the alternative approach which is commercially-led, and involves establishing a commercial incentive through creating a series of market opportunities, which would depend crucially upon the successful discovery and recovery of water ice from the polar regions.

The Lunar Commerce Portfolio, developed within the Moon Village Association (an NGO having Observer status at the United Nations Committee on the Peaceful Uses of Outer Space, in Vienna), is the mechanism which attempts to quantify these market sectors. This book is a primer for using the data, and the algorithms, within the Lunar Commerce Portfolio to try to understand whether the business case can be made to close for operating a Lunar economy. The Moon Village Association is working with the business school of Bocconi University in Milan, Italy to continue to develop the economic model and produce further versions, but this book focusses on the assumptions and findings of the first attempt, ie Version 1 of the Lunar Commerce Portfolio – thereby capturing the snapshot in time represented by this November 2022 Version 1.

"I'll put a girdle round the earth
  In forty minutes"
  – Shakespeare, A Midsummer Night's Dream

4

The Old Way

The last time we went to the Moon, with the NASA Apollo program, it cost the US taxpayers over 4% of GDP for the decade of the sixties. The current funding levels of NASA are nearer 0.4% of GDP (Budget of NASA, 2023). Furthermore, in surveys, the US general public have indicated that this level of funding is appropriate for the nation's civil space program. It seems that 0.4% is plenty (Pew Survey, 2023). And this is the background behind the current NASA plans regarding its proposed return to the Moon program, known as Artemis. Incidentally, that same Pew survey reports that sending astronauts back to the Moon should only be the eighth priority for NASA (following such items as climate change monitoring and asteroid protection).

There are a lot of parallels between the Apollo and the Artemis approach to getting to the Moon. Both of them use a single expendable rocket to transfer astronauts and their life support to the Moon. In the case of Apollo, it was the Saturn V (Saturn V, 2023), and for Artemis, it is the Space Launch System, SLS (SLS, 2023). As was the case with the Apollo Command Module, only the Orion crew capsule (Orion, 2023) is destined to return to Earth with its crew at the end of the Artemis mission. From the general area of lunar orbit, there is a lander which carries the crew (Human Landing System, HLS) (Human Landing System, 2023), and also potentially a lunar rover (Lunar Terrain Vehicle, 2022), to the lunar surface. The main stated rationale continues to be the furtherance of science and exploration. NASA owns and operates the rockets. All the life support for the crews will be provided with the initial delivery to the lunar surface.

Current funding levels for some of the international space agencies are listed in Fig. 4.1, and it is clear that NASA remains by far the biggest contributor.

D. Webber, *Lunar Commerce*, https://doi.org/10.1007/978-3-031-53421-8_4

SPACE AGENCY BUDGETS			
Country	Agency	Budget $B USD	Year
USA	NASA	23.5 B	2022
China	CNSA	11.7 B	2021
Europe	ESA	7.4 B	2020
France	CNES	3.5 B	2022
Russia	ROSKOSMOS	2.0 B	2022
India	ISRO	1.8 B	2022
Japan	JAXA	2.3 B	2022

Fig. 4.1 Comparison of main international space agency budgets. (Credit: Wikipedia, 2023)

There are some differences with Apollo, however. For instance, there will in Artemis be an orbiting Gateway space station in lunar orbit (Gateway, 2023), where crews enter and vacate the lander, whereas in Apollo the command module served this purpose. There are potentially some real benefits to having the Gateway station, especially with the potential of locally produced cheap fuel for missions either going onward further into the solar system, or indeed returning to Earth. It also might represent a great space hotel in lunar orbit for future lunar orbit space tourists. Also, this time, NASA is augmenting its resources by sharing the mission with providers in other countries, such as Canada, and in Europe. At this time (see Appendix C), 28 countries have signed up to be involved (Artemis Accords, 2023). Furthermore, this time round there is an explicit reference to the need to try out technologies that would eventually be needed for exploration on Mars. It is not intended for NASA to own the landers, as they had done with the LEM (LEM, 2023) during the Apollo era. Crucially missing this time, however, is the urgency of Kennedy's appeal. And so, there has been little incentive toward reducing costs in a cost-plus environment. There have been some gestures toward a more commercial approach, including use of GLXP-derived low-cost precursor landers and rovers. But this has not been enough of a change in the US government program to address the biggest cost item—the cost of getting to the vicinity of the Moon in the first place. The space agency is, however, awarding Artemis program contracts to providers who would offer the "lunar landing and return to lunar orbit" segments as a service, and who would retain ownership of the assets for future repeat delivery missions (both for the space agency and other potential future customers of the lander). The same approach seems to be the case with rovers (Lunar Terrain Vehicle, 2022).

However, with the current available funding levels, and those anticipated in the period up to about 2030, in the NASA and other agency budgets, operating in "the old way," it is only possible to mount a limited program of Moon landings. The current plan (Current Artemis Plan, 2023) involves only the following few missions:

Artemis 1 Launched 2022	Uncrewed lunar orbit and return
Artemis 2 Nov 2024	Crewed lunar fly-by
Artemis 3 Dec 2025	2-person lunar landing, via Starship
Artemis 4 Sept 2028	Crewed lunar orbit; building Gateway station (lunar orbit)
Artemis 5 Sept 2029	Crewed landing with lunar rover (lunar terrain vehicle)

Which takes us to the end of an early phase of the return to the Moon, as this decade ends. This outcome is a direct result of the choice to use expendable launch vehicles (SLS)—costing at least two billion dollars each (excluding development costs) for the mission. Using this "old way" single-sourced, expendable, cost-plus procurement approach produces costs to the taxpayer (you and me) which are at least an order of magnitude greater than alternatives. As we have discussed elsewhere (Webber, 2017), a firm-fixed competitive approach is to be preferred and, moreover, provides its own motivation and cost-control mechanism.

Notably missing from any of this is a detailed and clear plan about how the "sustainable" part of the enterprise is supposed to happen. Over many years, NASA has been providing funding (often via LSIC and LEAG) to dozens of enterprises who each independently analyze and test various parts of the potential lunar endeavor. There are piles of reports and test results. Each one it seems being followed by another stage of work. But lacking is a single agreed sense of direction, a single agreed anticipated architecture. It is a classic case of "analysis paralysis," where it is always possible for someone else to propose something else to test, and obtain funding to do so. But no one has the responsibility, it seems, to pull it all together as a unified plan leading to a permanent lunar settlement. And of course, the reason is clear. We just cannot afford to do it the old way. So, we keep on doing more and more tests and studies. We keep trying to design the optimum system. No one calls "enough!" And declares the way ahead. We have to keep learning more and more acronyms.

A consequence of this is a disconnect between approved funding levels and stated rhetoric about the intent of the program. It is not supposed to be merely a "flags and footprints" mission this time. The intention is declared to set up a "permanent presence" on the Moon, viewing our celestial neighbor as a resource, or even as a new starting point for future crewed missions to Mars (taking advantage of the much-reduced escape velocity from the Moon compared with from the Earth). NASA has recognized this imbalance between means and intentions (Foust, 2023) and has therefore sought help in closing the gap from businesses that could assist based solely on a truly commercial profit motive, rather than from direct government funding (which as we have indicated will be constrained at least in the period up to 2030).

Is there something wrong with the intent of the program, or with the funding levels, or both? Quite clearly, the existing approach, using the "old way," will not make it possible to arrive at a permanent human presence on the Moon at least up to about 2030. And this lack of adequate and steady government funding, furthermore, makes it difficult for possible new ventures to raise capital to attempt to provide the services, based on commercial operation, to fill in the capability gap between the NASA rhetoric and funding schedules. Venture capitalists are notoriously risk-averse, and certainly will be unwilling to step in where governments are unwilling to take the risks.

So, you will have realized by now that we are going to be needing a paradigm shift in how we undertake such missions, if we are ever going to be able to realize the benefits of embracing the resources of the solar system as part of humankind's prerogative. We need a new way.

References

Artemis Accords. (2023). NASA.gov/specials/Artemis-accords/index.html

Budget of NASA. (2023). Wikipedia.org/wiki/budget_of_NASA#

Current Artemis Plan. (2023). NASA.gov/specials/Artemis

Foust, J. (2023, March). NASA warns of devastating impacts of potential budget Cuts. *Space News*.

Gateway. (2023). NASA.Gov/Gateway/Overview

Human Landing System. (2023). NASA.gov/content/about-human-landing-system-development

LEM. (2023). *Lunar excursion module.* https://en.wikipedia.org/wiki/Apollo_lunar_module

Lunar Terrain Vehicle (LTV). (2022). NASA.gov/feature/nasa-makes-progress-with-new-lunar-terrain-vehicle-moon-rover-services

Orion Spacecraft. (2023). https://www.lockheedmartin.com/Orion

Pew Survey. (2023). *Americans' views of space: US role, NASA priorities, and impact of private companies*. Pew Research Center.

Saturn, V. (2023). https://historicspacecraft.com/rockets_saturn_5

SLS, Space Launch System. (2023). https://en.wikipedia.org/wiki/space_launch_system

Space Agency Budgets. (2023). Wikipedia.org/wiki/list_of_government_space_agencies

Webber. (2017). *No bucks, no buck rogers*. Curtis Press.

5

The New Way

The New Way is exemplified by two specific, and yet very different, enterprises, both of which have had a major impact on making a future lunar economy possible. Neither of them is a government entity. They have both played their part in making lunar activity more affordable and available for private access. They have both made possible orders-of-magnitude reductions in the costs of getting to, and operating on, the Moon. The two enterprises are Elon Musk's SpaceX (SpaceX, 2023) and the Google Lunar XPRIZE (GLXP, 2015). The what? Stay with me and an explanation follows. Let's look at the achievements and commonalities of approach between them.

What has SpaceX done? Elon Musk started it in 2002, at age 31, with the aim of making space launch more affordable. By December of 2003, SpaceX had designed and built its Falcon 1 launcher, and Musk even brought it to Washington DC, where I saw it parked right outside the National Air and Space Museum, and it made a direct challenge to all the legacy launch service providers (and indeed their governmental and military customers), announcing their availability to outcompete them. In typical Musk style (very representative of the New Way), the new launcher was displayed over night with blue light illuminations. It had been built by a small team of young persons. By 2007, it had successfully launched a payload into orbit—and this had been done with full access to the public via live video feed, and even including images from the vehicle showing ascent and payload separation. At that time, this was all very new. Musk's style embodies risk-taking and invokes a constant series of tests and learning experiences, sometimes involving catastrophic failures. But progress is the theme. After the success of Falcon 1, SpaceX began designing and building upgrades. By 2012, they were even launching payloads for NASA to the International Space Station using their upgraded

D. Webber, *Lunar Commerce*, https://doi.org/10.1007/978-3-031-53421-8_5

launcher and their new Dragon capsule. They then advanced to taking and returning human astronaut cargoes to the ISS, while simultaneously testing out and perfecting the technology for retrieving spent rocket stages from space for re-use. By 2015, SpaceX had demonstrated that they could reliably do that, and land the spent stages either at sea on a barge, or back at the launch site, on land. SpaceX brought us the technology of re-usability, with the implicit possibility of massive reductions in cost to orbit (or to the Moon, for that matter). It is worth stressing, for modern readers, that none of the legacy launch manufacturers (i.e., Boeing, Lockheed Martin, etc.), who had been receiving NASA contracts since the sixties, had managed to do this. It had not been in their interests. They were being paid for expendable launches on a cost-plus basis, with therefore guaranteed profits. But this of course had meant that (until SpaceX arrived on the scene) there was no incentive to reduce costs to orbit, or open up new commercial markets. It is not possible to overstate the importance of the SpaceX contribution.

What about this Google Lunar XPRIZE thing, already mentioned in Chap. 2? Surely, theirs cannot be compared to the achievements of SpaceX? I think they can. In their own way, they also moved the needle of affordable access to space, and specifically the Moon, significantly in the affordability direction. The XPRIZE Foundation, by means of its Ansari XPRIZE competition in 2004, had already by embracing risk and the use of prize money incentives, virtually created the suborbital space tourism market. Today's Virgin Galactic space tourism SpaceShipTwo is a direct descendant from SpaceShipOne which won the $10M Ansari XPRIZE in 2004. I can tell you that I was there at the time and was mightily impressed by the enormous crowds of young people—representing our collective future—who had turned out early in the morning to rendezvous in the middle of the Mojave Desert to support the attempt. Thus, were introduced the ideas of reusability in space access, low cost, and nongovernmental architectures. Moving on to 2013, I began to observe the activities of the Google Lunar XPRIZE, which was designed to make possible nongovernment and low-cost access to the Moon, by offering a $40M prize purse. The small international teams which set out to try to win the prize consisted generally of young people who had not even been born when Apollo was taking place. And they had a different approach. They used off-the-shelf Go-Pro cameras, for example. They used off-the-shelf software and PCs to design their flight profiles and navigation routines. They built small and negotiated shared space on launchers to bring down the launch cost component. The panel of nine volunteer international judges was able to offer guidance and awarded interim prizes for risk-reduction activities undertaken. The competition took longer than the funders (Google) had imagined, and so the competition was ended in March 2018 before any team was able to launch. Subsequently, two GLXP Teams have indeed launched their payloads to the lunar vicinity—Team SpaceIL from Israel, and Team Hakuto from Japan—although neither of them achieved a soft landing by the time of writing (Aug 2023). But nevertheless, the GLXP had been a very successful exercise and had made possible a new class of lunar vehicle many orders of magnitude less costly than before. Something that NASA was able to use to its advantage only a few years later when they created the CLPS contract opportunities. Indeed, the cost savings were so dramatic that NASA could even contemplate using several of the low-cost landers and rovers within the earlier cost regime of just a single "Old Way" lander or rover. So that there was room for failure without putting the whole project at risk. This, therefore, is I believe more than adequate justification for including GLXP alongside SpaceX as key enablers of the new phase of commercial lunar activity.

How do we put these pieces together to create a New Way to the Moon and its eventual commercialization? Let's take a longer view. Start thinking of the Moon as Earth's gateway to the solar system. Let's figure out how to make money on the Moon as a way to help fund the sustainability objective. Hint: It will *not* be possible using launch vehicles costing ten times the commercial rate. We are going to dig into this in detail in Part III, but one thing we shall learn from that work is that the "steady state" era when this becomes possible will not happen quickly. It might not be a self-sustaining economic system until 20 plus years into the future. It needs a special blend of governmental and private commercial entrepreneurial activities for which we have little experience, the ideally simultaneous development of a range of businesses which will share a collective and interconnected future. At least within the western democratic tradition of government, it will require a shared common purpose between commercial entities, national and intergovernmental agencies, a shared understanding of the respective roles of governments and businesses, and the support of taxpayers. Let's admit that it's not going to be easy. Of course, there will be intermediate stages, after 2030, when *some* of it may happen. Clearly, there will be ongoing science on the Moon, and contracts will be awarded to carry it out, but the paymasters for this will be governmental. We are going to study what are the likely true lunar commercial businesses and their interactions—both on the lunar surface and in lunar orbit—and make our best assessments of the likely timeframe, and the scale of the consolidated revenues that would result, if indeed we can succeed in bringing it about. This promises to be challenging, and very important, but fun nevertheless—the New Way.

We must bring together an understanding of technology levels, the potential geopolitical will, the potential market sectors expressed in terms of commercial competitive realities. Within the orbit of the United Nations, the Moon Village Association has been working on these matters, together with the implied regulatory environmental infrastructure, in order to support these kinds of initiatives. Specifically, under the auspices of its Lunar Commerce and Economics Working Group, this material was assembled together by a group of international volunteer analysts over a two-year period, in the aforementioned report and model known as the Lunar Commerce Portfolio (LCP, 2022). In doing this work, data was collected from firms, investors, and stakeholders, and incorporated into the narrative using carefully and rigorously defined interconnected market categories. Why? Because we need to understand whether enough revenues can be generated commercially to underpin the creation of the proposed permanent presence on the Moon. And if we can't do that, then to try to identify where the further work will be needed in order to close the business cases. The idea is to create a sound basis for continued improvement and recalibration and a focus for future research efforts. Indeed, measures are already in place to systematically do just that. The Moon Village Association signed a Memorandum of Understanding in November 2022 with the SEE Lab at the business school of Bocconi University, in Milan, Italy, to ensure that the work continues by using a Bocconi analytical team in helping the volunteer efforts of the MVA Working Group in improving the usefulness of the LCP (Bocconi, 2022).

So, in summary, the New Way requires that an effective trading estate of businesses can come together on the Moon, which will provide necessary goods and services for the permanent residents, using the available lunar resources without the need for delivery of supplies from Earth, and thereby generate enough revenues, when used in conjunction with government funds, to make the whole proposition affordable. That already sounds difficult enough, so why do we insist that there should be no assumed deliveries on an ongoing basis from Earth? Good question. Because such deliveries, it is assumed, would be very expensive compared with the equivalent locally-sourced alternatives, and furthermore, one of the stated objectives is to prepare for a possible future crewed mission to Mars, and in that case the journey times from Earth would make deliveries from Earth (with at least a six-month travel time) an unworkable solution. We are therefore deliberately giving ourselves an extra tough challenge. So, *in-situ* resource utilization will be essential (ISRU, 2023). We've simply got to learn to "live off the land." And, importantly, this time our geographic focus must be where the water is, hence the polar destinations (c.f. the generally equatorial landing sites of Apollo). The

Clementine (1994) and LCROSS (2009) missions found that there was water ice on the Moon, at the poles, in the permanently shadowed regions (PSRs). Let's explain that a little.

Over the majority of the Moon's surface, the environment varies according to the steady change from the intense cold of the lunar night (lasting two Earth weeks) to the high daylight temperatures throughout the two-week-long lunar day. These PSR regions exist because, unlike with the Earth, the Moon's axis of rotation is almost perpendicular to the plane of the solar system, and so the sunlight always reaches the lunar poles at grazing incidence (take your choice—up there its either always sunrise, or always sunset), and deep craters at the poles will therefore never see sunlight at any part of the lunar day. In other words, unlike Earth, there are effectively no seasons on the Moon. When you are in a PSR, you will never, ever, see sunlight (whereas at the North pole on Earth, depending on the time of year, you might not see the sun for months, but eventually there will be perpetual sunlight even through midnight). And this has been the case on the Moon for a very long time, so the volatiles have been able to collect there. This means that, at least in principle, we can make a rocket fuel factory on the Moon. And because the Moon is already far outside of Earth's gravity well, it makes an excellent way station for refueling vehicles that are heading further outward (just like gas stations on Earth situated just outside the city limits), or merely returning to Earth. Therefore, we potentially have a much cheaper source of rocket fuel for such ventures.

We need to figure out if these water ice resources, and other mineral content, when taking into consideration the vastly reduced costs of getting stuff into space now becoming possible in a genuine competitive commercial environment, will be enough to commercially sustain a human outpost on our celestial neighbor. Incidentally, the same lunar geometry ensures that, as well as Permanently Shadowed Regions at the lunar poles, there will also be some select corresponding pieces of lunar high ground which remain in perpetual sunlight throughout the lunar day and night. Great places to erect solar array farms. They are referred to as Peaks of Perpetual Sunlight (PPSs), and taken together, with the sun as a source providing power, this results in an architecture where we can live and mine throughout the lunar day, all year long. Figure 5.1 gives you an idea of what it would be like, capturing the elements of a lunar polar architecture developed by John Mankins within the Moon Village Association. The Earth remains visible just above the lunar horizon, and the sunlight is coming in at grazing incidence. In the image, you can see various habitats and antennas, and solar power collectors. I have to confess that it does not look much like living in Rio, say—but lots of folks manage to

Fig 5.1 Living and working on the Moon. (Credit: MVA_2021_Moon_Village_by_XTEND)

enjoy living on Earth beyond the Arctic Circles. And as we now know, we do have a serious purpose for doing this. Just so long as they have a neat bar and maybe a games room and a good communication link to the Internet and back to Earth.

Before we venture forth in the next section to figure out whether this is going to be do-able, it is worth laying out what is distinctively special with operating on the lunar surface, or in lunar orbit, and which will need to be taken into account when attempting to operate in the lunar vicinity. The conditions are not benign, and any commercial business depending upon lunar operating revenues will need to ensure that the technologies they will employ are capable of managing the conditions. Being able to operate at all is the basic requirement, before we can even consider whether we can make a buck out of offering the products or services to others at price levels that customers will be willing to pay. So, here are some lunar basics:

Gravity	1/6th g
Temperature	+250°F (+120 °C) to –210°F (–130 °C)
Vacuum	Negligible atmosphere
Dust	Lunar regolith is course and has electrostatic properties
Radiation	No atmosphere to protect against cosmic rays, etc.
Sunlight and Earthlight	One side of Moon always facing Earth. Rotation of Moon once/month so on the surface, there is sunlight for 2 Earth weeks at a time, then two Earth weeks of night. Some special "points of perpetual sunlight" in polar regions. Also, relatively nearby, there are some special cold traps which never receive any sunlight. There will always be some Earthlight over the entire nearside of the Moon.
Types of surface	Mare (lunar "seas") and mountains

It seems daunting, perhaps, but remember that we have done it before, and in the new lunar economy, there would be lots of opportunities to work and create a living in parts of the supply chain that does not even have to face the worst rigors of the lunar environment. You could be the proprietor of the first lunar hardware, supplies and service store, or trading post, for instance. Imagine it as a bit like operating the Hudson Bay Company store in the early days when Europeans came across the forbidding Atlantic to start a new life in the frozen far north of Canada or North America. Instead of trading furs for tools in the early days of the Europeanization of North America with the native Indians, on the Moon, the trading will involve selling supplies to the settlers, who will pay with whatever emerges as lunar currency. Maybe if you are a lunar water miner, you could provide water to the store, in return for some fancy food from Earth. Or maybe you need a newly refurbished helmet for your space suit.

This, then, is the new way. Increasingly, through time, the commercial service providers will fuel the growth, eventually overtaking the contribution from taxpayers on Earth. This is the route to sustainability. This is the beginning of mankind's economic development of space resources. This is how we go. If we can make it work.

References

Bocconi. (2022). SDABOCCONI.it/en/news/22/11/a-new-partnership-between-mva-and-sda-bocconi-school-of-management

GLXP. (2015, November). *The Google lunar Xprize-past, present and future*. In Barton, Wong and Webber Reinventing space conference, Oxford.

ISRU. (2023). *In situ resource utilization*. https://www.NASA.gov/ISRU

LCP. (2022). Moonvillageassociation.org/download/the-lunar-commerce-portfolio-first-edition-november-2022

SpaceX. (2023). https://www.spacex.com

Part III

Paying Our Way

Having decided to pursue the non-traditional and commercial approach to returning to the Moon on a sustainable basis, this section lays out the series of detailed assumptions which were made in the course of creating the Lunar Commerce Portfolio, using the best data that could be found in the public domain in the two-year period leading up to the release of Version 1 of the Lunar Commerce Portfolio in November 2022. The object of the exercise is to try to estimate whether or not, and on what timescale, it will prove possible to pay our way into operating a permanent lunar establishment.

Initially eleven inter-related market sectors were established to represent the elements of a possible self-sustaining commercial lunar trading estate.

Then, for each of the identified sectors, a database was created involving precise sector descriptions, and the potential suppliers and customers, value chains, drivers and constraints, and potential price assumptions.

This information was then fed into a demand model, which eventually produced a first-cut at possible lunar revenue generation.

The model also makes clear the degree of uncertainty, and therefore risk, involved at this early stage of estimating lunar commerce revenue potential, and underlines those areas where more work is required to fuel subsequent versions of the Lunar Commerce Portfolio. Also identified are those extraneous factors which would be expected to impact the revenue outcomes, several of which involve the consequences of possible international and national regulations.

"There are more things in heaven and earth, Horatio,
*   than are dreamt of in your philosophy"*
*   – Shakespeare, Hamlet*

6

Fundamentals

We pay our way into becoming a space-faring species by engaging in space commerce. And we have seen how *lunar* commerce is the necessary element in getting started on that ultimate long journey of mankind. Now we begin to assess what exactly are the lunar markets? And who is going to be selling what to whom? In other words, we must establish the lunar market fundamentals. How are we going to begin to assess what are the businesses which will ultimately help us pay our way? What is the model? The first key step is to develop a view of the likely *potential lunar market sectors*, how they interrelate, and ensure that the list is complete, and that it avoids double-counting of activities.

There will of course be the ongoing activities related to pure science and exploration objectives, as were conducted 50 years ago during the Apollo missions. For example, governments may in future choose to fund the deployment of a large array radio telescope on the lunar farside. We did not in the LCP treat the aggregation of such potential government activities as a true market segment in its own right, but instead represented the associated work through allocated contracts across a number of impacted commercial market sectors, represented by potential government funded contracts in, e.g., transportation, facilities, and maybe manufacturing sectors (thereby avoiding double-counting).

After a great deal of discussion, the consensus established among the analysts creating the Lunar Commerce Portfolio resulted at that time in nine separate sectors, many of which in themselves contained subsectors (Tiwana & Webber, 2021). Some subsectors were grouped together merely for administrative convenience (e.g., shortage of volunteer analysts) and therefore did not necessarily represent the most logical, or best, way to record the array of lunar commercial activities. So, subsequent to the issue of the LCP in

November 2022, and for this primer book, a few minor changes have been introduced within the same overall envelope of market demand, to facilitate explanation and scheduling aspects, and the resulting set of market sectors of the lunar economy is now represented by the 11 segments shown in Fig. 6.1. A few of these market sectors have counterparts in lunar orbit, as well as those on the lunar surface.

We used precise language to carefully describe each sector to make sure that there were clear distinguishing elements in any potential overlap areas. And the resulting market sector fundamentals will be presented in turn in the subsequent Chaps. 7 thru 10 below. We mentioned earlier about the case of science and exploration, and how they have been treated. But maybe we should use another example to make sure we understand what we mean by *double-counting*. This means we must make an important diversion to discuss lunar tourism.

Lunar Tourism

An example of a potential overlap area would be lunar tourism, for instance. It wasn't, as you can see from the diagram, given a separate sector of its own in the LCP, but instead it was dealt with by making sure that each of its constituent parts was being handled systematically across the various other sectors—e.g., there would need to be the associated allocation within the sectors for launch vehicles, landers, rovers, hotels, communications, power, food, etc. Because of the importance of lunar tourism forecasts to the whole of lunar commerce, we used a standard set of assumptions, and then parsed them out among the other sectors. The original analysis is provided in Appendix A, where the methodology, market research data, and reference sources are set out. In fact (Spoiler alert!), the estimates of the market potential of lunar space tourism turn out to be simultaneously the least well understood, yet the most needed, in generating lunar business in almost every sector. In summary, the following were the lunar space tourism assumptions used for this work (with obvious implicit wide ranges of uncertainty), and quite clearly, there is a need for some statistically valid demand-based market research among very wealthy people to understand their level of interest in the proposition, and to establish likely price levels. Notice that this is an example where a lunar orbit counterpart of the sector is included. We look therefore at both lunar surface and lunar orbit tourism demand:

For lunar orbit tourism: from 5 to 130 tourists at a time staying for two weeks, depending on price.

For lunar surface tourism: from 1 to 10 tourists at a time staying for two weeks, depending on price.

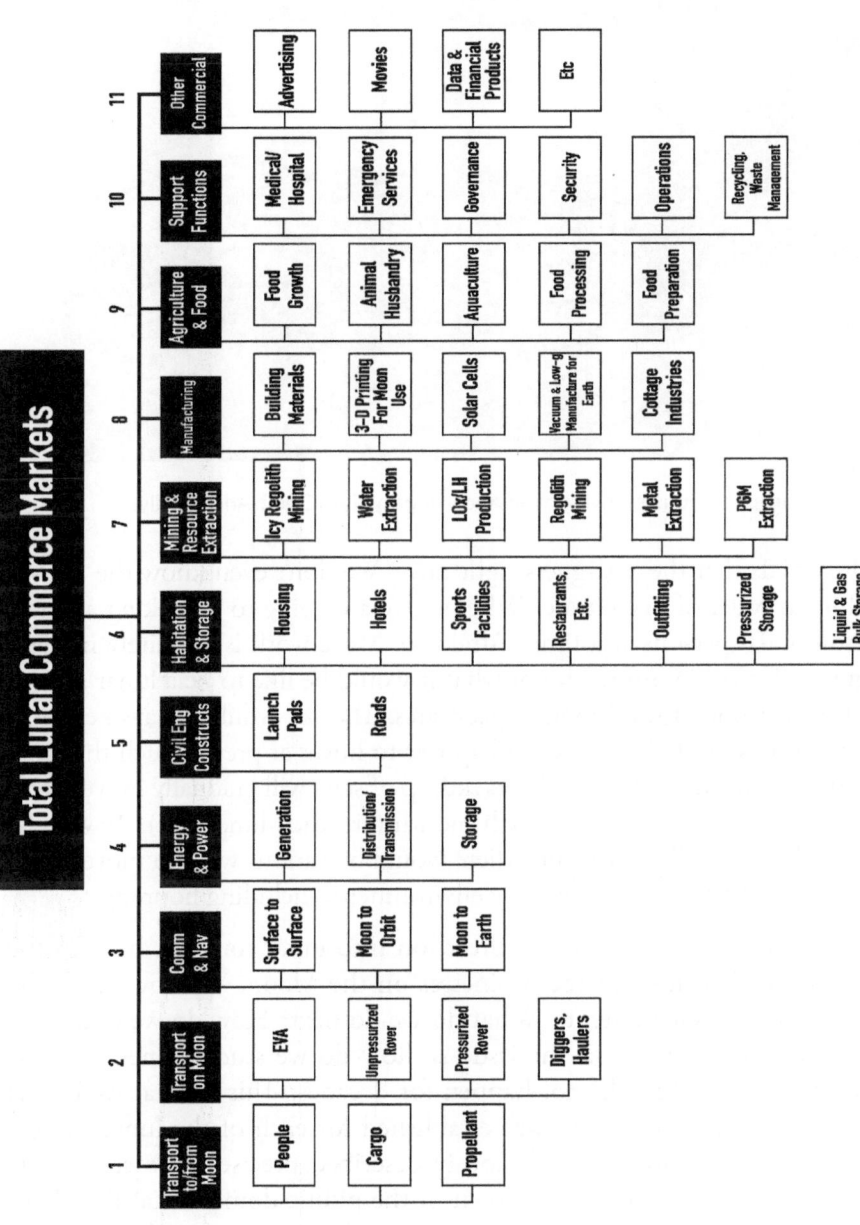

Fig. 6.1 The array of commercial market sectors and subsectors which, when taken together, would constitute the true lunar economy. Note that some markets, such as lunar tourism, are not separately identified, but span several sectors. (Credit: DW/MVA)

Fig. 6.2 Lunar space tourists arriving at an Apollo legacy site. (Credit: Arlene Kelly)

Really? Is that the best-guess right now? We don't even know the answer within an order of magnitude? Whether we are going to be talking about 1 person or 10 persons at a time? Afraid so. More work is definitely needed. Figure 6.2 provides some idea of what it would be like to be a lunar surface tourist near one of the Moon's legacy sites. The sky would always be black, even at mid-day. The Earth would appear to hover at pretty much the same location in the lunar sky, but the backdrop of stars will gradually move across the sky in the course of each Earth month (viz. each lunar day). As was the case with the Apollo astronauts, there would be various ways to move about to adapt to the Moon's low-gravity environment, including hopping.

Now, after that very important diversion into lunar tourism, back to the mainstream of figuring out the businesses on the Moon. We now know the segments and the subsegments. What do we do next? How do we build this first portfolio of businesses? The issue is, how do we study something that doesn't exist yet, and might not happen for decades? This is what we did. *A standard data collection routine* was established for each of the lunar market sectors, and this involved the previously-described precise *sector description*, the potential *suppliers* who have shown in the public domain that they have an interest, the potential *customers* for the product or service, the interconnected supply or *value chains*, and the *drivers and constraints* which would impact development of the particular sector. Drivers would include setting of price levels, and constraints would include supply chain challenges for the sector. This is, in effect, the portfolio of possibilities.

Table 6.1 Common driving assumptions for the Lunar Commerce Portfolio, Version 1

Common driving assumptions		
Assumption	Early phase	Mature phase
Annual crewed missions	1–9	1–40
Annual cargo missions	0	1–40
CLPS-like missions/year	2.5	2.5
Resident people on surface	2–4	40–120
Resident people in orbit	0–2	0
Number of bases	1	2–4
Lunar orbit tourists	5	100–3200
Lunar surface tourists	0	10–125

Credit: DW/MVA

It was then recognized that the sector demand, and therefore revenue potential, would vary as a consequence of a range of *key driving assumptions*, common across all sectors. They were developed by consensus of the team of international analysts working on this, taking into account *inter alia* the lunar tourism analysis provided in Appendix A, and are listed in Table 6.1.

This is an important table. Clearly, it contains assumptions with enormous ranges of values. This was the best that we could do, given the inherent uncertainties on our future lunar ventures. We have put our trust in the previously noted dictum of polymath Francis Bacon (1560–1626), whom you may recall told us: "If a man will begin with certainties, he shall end with doubts, but if he will be content to begin with doubts he shall end in certainties." This table effectively shows our initial doubts. This is, in effect, our reality check. And these variations are addressed later in Chap. 12. But it is also important fundamentally because of what it means about our demand-based approach in the Lunar Commerce Portfolio. Certainly, the work of the LCP is demand-based, rather than being fueled by manufacturer-hype, yet it is nevertheless driven, in this Version 1, by these stated assumptions. If you don't agree with them, at least you know what they are, so you can make changes. So, we can already see that it's not going to be possible to come up with one nice, clean, number to represent lunar commerce business potential. At least we have ensured that the same assumptions were used for each of the 11 market sectors, ensuring consistency.

The assumptions were the result of a careful analysis, taking into account all that could be gleaned from public sources, but could still of course be very wrong. We could point out, for instance, that since the November 2022 issue of the LCP, there has been an emerging belief that maybe the expanding use of AI-driven robotics will come to dominate the economic development of the lunar economy. And if this would indeed come to pass, it would have a major impact upon the results of the business analysis presented here. There would be much less need for the human-oriented infrastructure. The good news, however, is that the models allow for changing these values for future updates of the LCP. And, indeed, this will be undertaken systematically by the Moon Village Association's Working Group, by the Bocconi University analysts, and by inputs from the newly created Lunar Commerce User Group (LCUG, 2022). We identify later what needs to be done to improve these assumptions, and narrow the ranges of uncertainty, but in general, it requires some systematic and statistically valid market research about demand, the publishing of experimental results about the first lunar resource extraction experiments, and some agreed consensus across national space agencies about the overall approach and commitment to this return to the Moon. You see how helpful it can be once you know where you are going, and why?

Beyond this, one further structural element needed to be added to the model of the Lunar Commerce Portfolio. We know intuitively that there can be an array of *external factors*, many of them geopolitical, that could impact the outcome. This was handled in the Lunar Commerce Portfolio by means of a *scenario treatment*. About 21 external factors were identified (see Appendix B), each of which could impact the revenue generation across all sectors, grouped in various ways to reflect a likely geopolitical reality. For Version 1 of the LCP, we considered four such scenarios (Alpha, Beta, Gamma and Delta), and Fig. 6.3 shows two of them, by example, to indicate how this was done. It is expected that this aspect of the LCP will be particularly useful to national and international space agencies, as they plan ahead. Certainly, the Bocconi University analysts intend to explore how use of the model can contribute to understanding international strategies, such as those that might emerge from the United Nations Committee on the Peaceful Use of Outer Space (UNCOPUOS) for regulating lunar activities (UNCOPUOS, 2023).

At this early stage of knowledge about lunar commerce markets, and the factors which could influence the revenue derivations, it was clear that it did not make sense to attempt a year-by-year revenue forecast. We would all be kidding ourselves. So, instead, the best assessments of *average annual revenues* for each of two distinct time periods were developed. The first of these time periods was named the *"Early Phase,"* and that was considered to be the period until 2030. For the second period, the likely average annual revenues were developed for a much later period, when sustainability had been achieved by use of ISRU, the lunar businesses had become established, and therefore when the sectors could treat each other as customers. This of course implied the need for the ideally simultaneous advancement in all 11 sectors. This period was called the *"Mature Phase."*

When would this Mature Phase take place? We honestly do not know. So, the analysts did not use a calendar date to stipulate the conditions, but instead agreed *a definition*. The Mature Phase was defined as *"the period when there is*

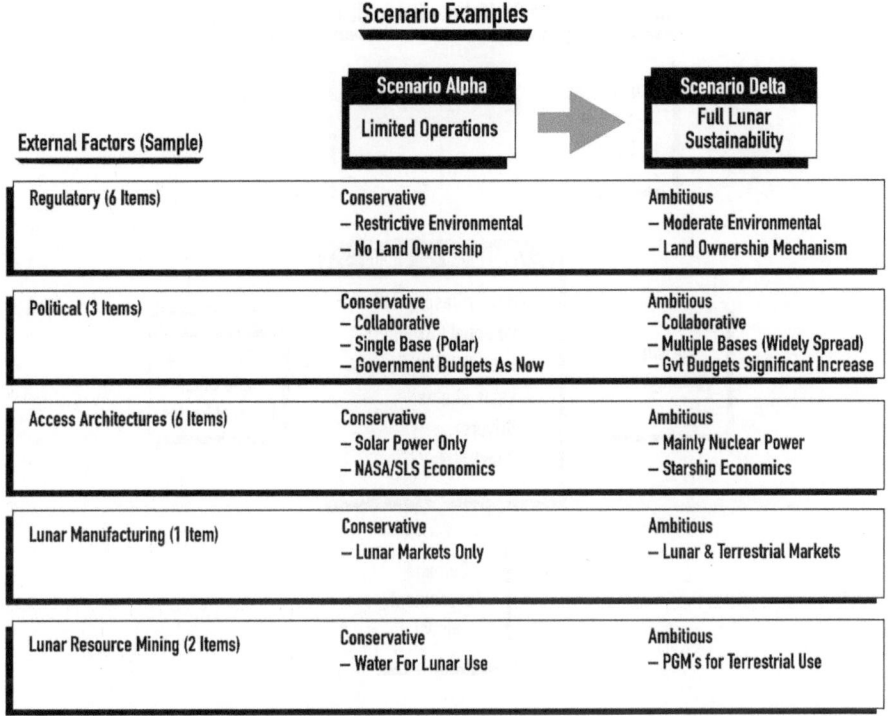

Fig. 6.3 Extraneous factors can affect lunar business results. The Lunar Commerce Portfolio studied these impacts via the creation of Scenarios of potential futures. This chart summarizes two of them, as examples. (Credit: DW/MVA)

a permanent human presence on the Moon sustained by the Moon's resources, and not dependent for the necessities of life on a logistical supply chain of deliveries from Earth." Clearly, there will likely be a gap of unknown duration between the results of the "Early Phase" and those of the "Mature Phase." And we did not worry about that. With more knowledge in future years, it will for sure become possible to improve these simple initial assumptions, and begin to remove some elements of uncertainty, and even perhaps to fill in some of the gap between the Early and Mature Phases. But at present, it is appropriate to acknowledge the limitations of the assumptions. In fact, it is an important message to government and other space planners that we are currently extremely limited in our ability to provide well-based revenue expectations for the era of lunar commerce.

With these steps, then, was created a model, and a vocabulary, or taxonomy, was established to make it possible to continue to develop and refine our knowledge of the lunar commerce potential. The Moon Village Association has, with its Lunar Commerce Portfolio, put in place (and incidentally made

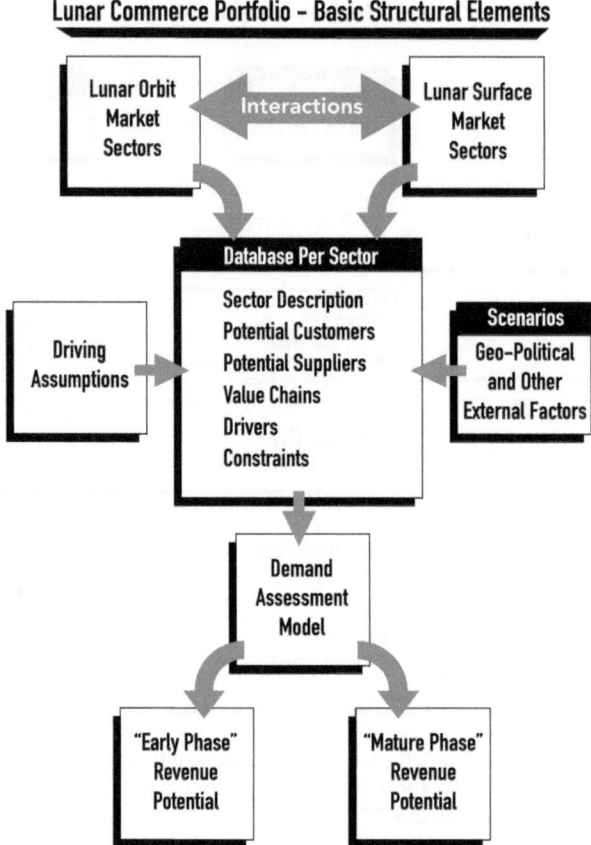

Fig. 6.4 Basic elements of the Moon Village Association's Lunar commerce portfolio report and model. (Credit: DW/MVA)

freely available) a means whereby potential investors can judge their opportunities and level of interest. They have made available the mechanism for ongoing improvement, using the MoU with Bocconi University's business school (Bocconi, 2022). And they set in motion the creation of an independent Lunar Commerce User Group (LCUG, 2022), who can collectively challenge the assumptions behind the LCP and its models. We can now all assess the value-added by each part of the lunar economy under differing circumstances. We can find out whether it can be made to work. That model is shown in Fig. 6.4. Let's use it and find out what the future might bring, as we attempt to create a lunar economy.

To do this, we proceed to discuss each of the 11 market sectors in turn, providing our best available current knowledge, as incorporated in (our slightly modified version of) the Lunar Commerce Portfolio, Version 1. For each of the sectors, we shall include a summary chart to the same format, making possible a ready means of comparing the sectors. You have stuck with us so far—now you will begin to discover "the meat" of our findings. This, indeed, will provide the detailed answers to the questions posed at the outset about "What will people be doing up there?" It may seem a bit "heavy duty," and/or even a bit repetitive, but we felt that the work was important to do, and furthermore important to record and make available. It is our contribution to "making it happen." This primer book, it is hoped, will be used as a future resource and reference for any given market sector, and so we have made the relevant information self-contained within each sector. It is of course important to understand the big picture. But it's also important to critique the individual market areas within the Portfolio. So, hang in there. We need you to follow, so that you can challenge any assumptions where you have better, or more recent, data. Of course, you may still feel inclined to keep in reserve the question about "Should they be allowed to do it at all?"—and we shall indeed come back to that in Part IV. But, essentially, we are now in a position to be able to assess whether, based on our declared assumptions, a lunar economy is indeed possible.

References

Bocconi. (2022). SDABOCCONI.it/en/news/22/11/a-new-partnership-between-mva-and-sda-bocconi-school-of-management
LCUG. (2022). Lunar Commerce User Group. https://lunarcug.com
Tiwana, J, & Webber, D. (2021). *Initial planning paper – Working group on lunar commerceandeconomics*.moonvillageassociation.org/download/lunar-commerce-and-economics-working-group-lce-initial-planning-paper-january-2021/
UNCOPUOS. (2023, March). *Technical presentation on the lunar commerce portfolio report*. Gidon Gautel.

7

Getting There

This is in many ways the most obvious first step to answering the question about who will be selling what to whom on the Moon. There will be the commercial business of someone paying someone else for the means of getting there in the first place. And for repeat trips for both people and goods. The category in the market sector descriptions (Fig. 6.1) of the LCP is sector 1— Transportation to/from the Moon, Launch Providers, and Lunar Landers. We are considering both the means of getting to lunar orbit, and from there to the lunar surface. We are considering both deliveries of cargos and of humans (and eventually some animals—eggs for breakfast, anyone?). And, of course, this is a two-way street. Sometimes, humans and cargos will be returning from the Moon back to Earth. Or simply from the lunar surface to lunar orbit. A component of the Artemis program is a Lunar Gateway Space Station in lunar orbit (Lunar Gateway, 2023). This market sector 1, therefore, is probably the best understood of the 11 market segments at this stage.

Technologically, it makes sense to consider in turn the different classes of vehicles needed for different purposes. There will need to be launch vehicles which can deliver a payload from Earth to lunar orbit (and sometimes even that step might involve more than one type of vehicle). The payload itself, whether human or cargo, will need to be carried within a spacecraft atop the launcher. And there will need to be the means of getting from lunar orbit (i.e., at the lunar gateway station) down to the lunar surface. And the vehicle which conducts that part of the mission is called the lunar lander. And generally speaking, the lander needs to have the capability to take off again to reach at least lunar orbit. Although, in some cases, the lander might remain on the surface to serve as a "habitat," or even be able to relocate, as a hopper, across the surface (as was proposed by the GLXP team Moon Express—and observed

D. Webber, *Lunar Commerce*, https://doi.org/10.1007/978-3-031-53421-8_7

by GLXP judges during tests). There could also emerge a class of simplified shuttle spacecraft which is designed simply to operate between lunar orbit and Earth orbit, not needing any of the complexities and weight associated with re-entry into Earth's atmosphere and gravity well, or of landing on the Moon.

From a financial standpoint, some of these craft will be bought and sold as a taxi service, with ownership remaining with the service provider. That will particularly be the case when a spacecraft is reusable. That will probably apply generally to lunar landers, and Earth–Moon shuttle craft. Sometimes the customer for the product or service will be the government; sometimes it will be a commercial entity. The trend over time will be away from the government as customer, and toward a commercial transaction on both sides, i.e., as provider and customer of the service. However, it is almost a certainty that there will not develop a true lunar commerce economy unless there is possible a massive shift in the economics of getting into space, and specifically in getting to the Moon. The attempt to make this change in launcher economics is taking place, both with regard to getting into Earth orbit, and to the Moon, as this book is being written. And the key is reusability. As discussed in Chap. 5, SpaceX is leading the way in this respect.

At present, the primary delivery system intended to transport both cargo and humans to the lunar surface as part of the US-led Artemis Program, in the period up to 2030, is the US government owned and operated Space Launch System (SLS), which is a very expensive proposition (SLS, 2020). It is always something of a mystery to obtain and understand the exact costs of NASA launchers. At least in part that is because it varies with the expected launch rate. But also, because they are contracted on a cost-plus basis, which encourages escalating costs. Estimates vary, but it seems that each SLS mission will cost about $2 B (yes, billions), excluding the development costs of probably more than about $23 B. It is indeed a remarkable thing that, a full half-century after the Apollo flights of the Saturn V vehicle, and given all the space launches since then, that NASA was not able to come up with a design approach for this new generation of Moon-launcher which included all the advancements that had been introduced by the commercial launch business since those early days. Nowadays, building launch vehicles has become quite routine. With modern developments in commercialization of both vehicles and home-built spacecraft (e.g., cubesats), it is nowadays possible (as demonstrated by the GLXP) for a high school class to build and afford to launch their own satellites, via ride-sharing arrangements.

Alternative commercial delivery systems being developed, and some already operational, involve reusability, which therefore offers the possibility of massive orders-of-magnitude reductions in price to Earth orbit and to lunar surface. This will be essential to the creation of the proposed commercial lunar economy. For example, estimates for the trip costs of the SpaceX Starship are around $100 M each (some say the figure could become as low as $2 M with experience and full reusability). There are also other commercial contenders which will also offer prices to the lunar vicinity at a tiny fraction of the SLS approach. For full detail of the alternatives, go to the Annexes of the Lunar Commerce Portfolio (LCP, 2022). So, let's now look at the assumptions we made for the vehicles designed to get cargo and people from lunar orbit to the lunar surface (and sometimes back again), the lunar landers.

The first contracts for landing systems for both humans and cargo are just being written as we write. There are a series of contracts, named the Commercial Lunar Payload Services (CLPS) program awards (CLPS, 2018), which are designed for getting relatively small robotic landers down to the lunar surface. They do not use the SLS. They have to find their own (commercial) way to the Moon. And then there are the human landing system alternatives being developed (at present with contracts to two commercial providers, Blue Origin and SpaceX).

The series of CLPS missions, which will take place from now until around 2030, are for robotic missions to the lunar surface, and many of them are being provided by the entrepreneurial companies created in attempts to win the Google Lunar XPRIZE (GLXP, 2015), which was indeed designed to bring about a low-cost solution and private access to the lunar surface. The objective was to build a spacecraft, effectively without government funding, and have it land on the Moon, travel for 500 m, and take and send high-definition images back to Earth. These GLXP craft were originally designed and built by "guys in garages" in 16 teams around the world and hence are a low-cost solution. All of the contracted CLPS missions include both the arrival on the lunar surface by a lander sent directly from Earth as part of a commercial cargo launch from Earth, and the subsequent deployment of a surface rover, or hopper (to be discussed under Chap. 8 below). In the GLXP, there was a total of $40 M of prize money available, with only $20 M for the Grand Prize, and lesser amounts for additional achievements, such as surviving the lunar night, or finding water, or even for photographing a lunar heritage site. The US space agency NASA is paying for these CLPS robotic lunar services via fixed price service contracts, in a change of philosophy which trades the old NASA way (owned, designed, expensive cost-plus contracting) versus a new way (buying only a service, low cost, high risk). In principle, it becomes possible to afford many such CLPS commercial missions for the price of a single NASA mission conceived and operated in the old way. These GLXP spacecraft, both landers and rovers, were very low-cost solutions. They had to be. Because they were originally competing for prizes that only amounted to $40 M in total, and they were not allowed by competition rules to secure any governmental funding (beyond a maximum of 10%).

The GLXP teams cut costs by using off-the-shelf equipment, such as Go-Pro cameras for imaging, and also by employing highly efficient trajectories, which needed relatively low fuel requirements, even though they took a long time to reach the lunar vicinity. In the course of the competition, and in pursuit of some interim milestone prizes proposed by the judges (designed to help with developments in landing, mobility, and imaging areas respectively), the several international teams learned how to reduce their risks by addressing problems such as how to handle lunar dust. The judging team attended demonstrations and development tests in the USA, India, Israel, Japan, Germany, and elsewhere, to observe and monitor how the young enthusiastic engineers were approaching the problems. There were 22 technical reviews in six different countries presided over by the GLXP judges. The team members learned how to navigate both *en route* to the Moon, and then subsequently on the surface. They needed to be able to accurately demonstrate that their rovers could arrive safely at the Moon, avoiding the Heritage Sites (which was a GLXP competition requirement introduced by the judges, following discussions with the Smithsonian Museum). And then, that they had moved 500 m, in order to claim the prize money (in the absence, of course, of the familiar terrestrial GPS system support). The judges were able to compare the results of several alternative approaches to the lunar navigation problem, and thereby decide on a compensation factor to be applied to the future lunar telemetry readings, in order to be able to certify the potential award for distance traveled. In the event, a decade after the competition had been conceived and announced, it ended in 2018 before any of the teams succeeded in winning the Grand Prize. But NASA realized that the competition had nevertheless ensured that there was a ready pool of candidates for meeting their CLPS goals. Three teams (Astrobotic and Moon Express from the USA, and Team Indus from India) each did, however, receive $1 M interim GLXP landing milestone prizes for demonstrating their advanced readiness level.

Two of the former GLXP teams have already made attempts at Moon landings, and have reached the lunar vicinity (Team SpaceIL, and Team Hakuto), but have not at the time of writing achieved soft landings. SpaceIL did make a hard landing in April 2019 after launching in February 2019. Team Hakuto (working with iSpace) had the same result in April 2023 after launching in December, 2022. Others of the old former GLXP teams have now reconfigured, have received CLPS funding from NASA, and so are now getting a second chance to land and perform on the Moon. The young engineers deserve success for all their efforts, starting with their engagement with the Google Lunar XPRIZE process as early as 2007. The many GLXP team members enjoyed the opportunity to meet with other international team members at various Team Summit events, compare notes, and continue to make their dreams of private access to the Moon a possibility and indeed a reality.

We should make specific reference again to lunar space tourism. Under our assumptions, lunar tourists will likely be a significant part of the "cargo" for this market sector. This will emerge in two forms—both as lunar orbit (or vicinity) experiences, and as lunar surface operations. There have already been contracts signed for lunar orbit (or vicinity) space tourism experiences (see Appendix A). Both of the contracts have been signed with SpaceX. One was signed by a Japanese billionaire Yusaku Maezawa in September 2018, and he intends to take some of his artist friends along for the ride. His project is called "Dear Moon," and yes, you can Google it. Subsequently, in October, 2022, another contract for lunar orbit tourism was signed, this one by former Earth-orbital space tourist Dennis Tito, who intends to take his wife Akiko with him on a circumlunar mission. Fly me to the Moon, indeed. At present, there are no deals yet for lunar surface tourism, and indeed, there are as yet no lunar tourism hotels in the design phase. However, market surveys (Adventurers' Survey, 2006) indicate that there will be demand for such offerings, once the service can be provided, and prices established. The terrestrial-based suborbital space tourism experience has now become operational—and indeed one of the earliest Blue Origin missions using New Shepard carried Sara Sabry, who was our LCP Team Lead for sector 2! We shall discuss these other aspects of lunar space tourism (habitations, food, etc.) in the appropriate sections below, but in the current section we are including the necessary launch provider and lunar lander services to support the expected demand. Indeed, the expected number of humans arriving at the lunar vicinity for lunar tourism purposes is an important indicator of lunar commercial activities in general, and so is included in the list of key driving assumptions discussed earlier. As a reminder, there is a detailed discussion and assessment of lunar tourism demand in Appendix A.

Figure 7.1 summarizes the Sector 1 dataset. With regard to the landers designed to deliver humans to the lunar surface, there are currently two teams under contract. The first one is SpaceX, which is contracted to provide the lander for Artemis 3, the mission which is intended to return humans to the Moon for the first time since 1972 (Apollo 17). Then, a consortium led by Blue Origin has also received a contract to perform the same service for NASA, on a subsequent Artemis mission. These two proposed human landers are shown in Figs 7.2 and 7.3.

Market Sector #1 Transport To/From Moon

Description   Delivery of People, Cargo, Supplies
To Lunar Orbit And Lunar Surface. Return Trip From Lunar
Surface & Orbit To Earth With People And Resources

Potential Customers   All 10 Other Market Sectors
Government Space Agencies, Military Universities,
Companies Incl Research Institutions, Non-Profits, NGO's,
Commercial Entities & High Net Worth Individuals

Potential Suppliers
Space-X, Boeing, NMA, ULA, Astrobotic, Dynetics,
Blue Origin, etc.

Drivers And Constraints
Trip Frequencies, Cargo Mass Capacities, Price/Kg,
Seats/System, Max Payload, Propellant Availability, etc.

Value Chain Schematic

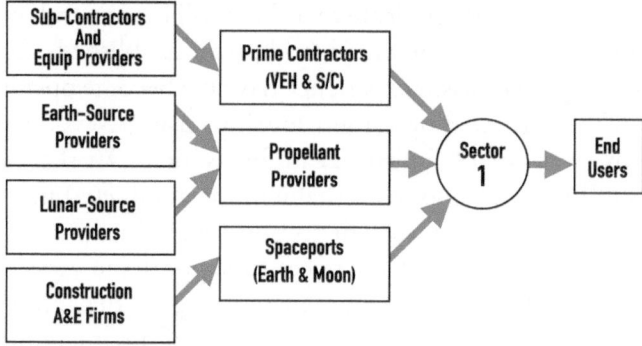

Fig. 7.1   Summary of LCP data for sector 1—transportation to/from the Moon. (Credit: DW/MVA)

Let's now be methodical, and consider systematically what is the status and expectation for this market segment # 1—transport to/from the Moon, as we begin this first quantification effort of lunar commerce as represented by the findings of the Lunar Commerce Portfolio, Version 1. The format we shall use is the best we could come up with right now so that, at our present state of knowledge, we could compare the market across all sectors, and ultimately make possible investment portfolio decisions. We will stick to this standardized setup (i.e., market, suppliers, customers, drivers/constraints, and value

Fig. 7.2 Proposed SpaceX Artemis Lunar lander. (Credit: SpaceX)

Fig. 7.3 Proposed Lunar Lander by Blue Origin consortium. (Credit: NASA/Blue Origin)

chains) throughout all of the 11 market sectors in this primer. I figured that you unfortunately must embrace this somewhat repetitive approach in order to be able to access the level of detail in the findings that you justifiably need. Forgive me if, on this point, I reckoned that providing you with the information is more important than the literary merits, or otherwise, of the exposition, accepting the downside. So, hang in there. You will find that even if the format is repetitive, the content will be enlightening and even surprising. And if you subsequently just want to make reference to one particular market segment, then you will find it a benefit to be able to find all the associated material complete and self-contained. So, here, as the first example of our standard information format, is the data portfolio for market segment 1—Transport to/from the Moon.

What Is the Formal Description of the Segment? The definition is "The movement of people, cargo, propellant, etc., between the Earth and the Moon and lunar vicinity."

Who Are the Potential Suppliers? During the data collection of the Lunar Commerce Portfolio, Version 1, 57 potential cis-lunar transportation service providers were identified, clearly with a considerable variation in readiness, with full details available in Annex B of the Excel model which accompanies the Lunar Commerce Portfolio (LCP, 2022). In the Early Phase, cargos are assumed to be delivered by CLPS landers (Astrobotic, Deep Space Systems, Firefly Aerospace and Orbit Beyond), and also possibly by the ESA lander. People are assumed to be delivered by combinations of Artemis providers (launched on NASA's SLS vehicle in the Orion capsule (Orion, 2020), then landing via the SpaceX or Blue Origin, or maybe Dynetics, lander). During the Mature Phase, many more providers may emerge, including providers from China and India. When we consider the full scope of this market sector, we need to include the terrestrial spaceport part. There are existing spaceports around the world, plus several being contemplated. A new spaceport creates opportunities for construction and A & E firms. Suppliers are particularly interested in the possibilities of lunar tourism opportunities, because this can result in special needs which suppliers can address. In this first issue of the Lunar Commerce Portfolio, some 15 Earth Launch and Recovery Sites (ELRS) were identified in 8 different countries (China, India, Japan, Europe/Guiana, Russia, Kazakhstan, New Zealand, and the USA). In addition, two companies were identified who were addressing the building of the spaceport launch and landing sites on the Moon. There are about 22 companies aiming to provide commercial lunar cargo lander services, in addition to four or five governmental potential suppliers (China, India, USA, Russia, and ESA). With regard to the crewed launch and lunar landings, at present 4 countries are making plans to provide capability (Canada, Russia, USA, and China), and one or two commercial firms are also lined up. And of course, the mainstay of this sector is the basic launch vehicle supplier, of which eight commercial entities in three countries were identified, augmented by four governmental providers (China, India, Russia, and the USA). Incidentally, in some architectures using Starship, it would be possible to conduct the entire mission from launch on Earth to landing humans on the Moon using the same vehicle.

Who Are the Potential Customers? For this sector, two tiers of customers were identified. In the Tier 1 category, the customers need the transportation services to pursue science, space exploration, national security and commercial venture development. In this category, there would be space agencies, governments, military, universities, research institutions, NGOs, High Net Worth individuals, and commercial entities. The Tier 2 category is reflected in all the other market segments from 2 thru 11, and represents service providers to the Tier 1 customers. All the rest of lunar business depends at least initially on this sector to get them started. Among the governmental customers are all signers of the Artemis Accords (discussed later). At present (a reminder that this means August 2023), this list consists of 28 separate states. Then there are the academic research institutions and commercial operators with payloads on the early CLPS missions (CLPS, 2018)—about 15 were identified in LCP Version 1. Examples of such already-contracted customer missions include Embry-Riddle University (USA) flying a cubesat camera system on Astrobotic Mission 1, University of Colorado (USA) with a low-frequency radio spectrometer on Intuitive Machines IM-1 mission, The Arch Mission Foundation, a nonprofit flying a payload to demonstrate the ability to host lunar archives of humanity's heritage, and Celestis, a space burial company with a payload on Astrobotic Mission 1. The Hungarian company Puli Space Technologies also has a payload scheduled for Astrobotic Mission 1. Incidentally, Puli was a small startup during the GLXP, and demonstrated an early rover prototype to GLXP judges during mission reviews in Budapest in 2014.

What Are the Likely Drivers and Constraints? Drivers of demand for this sector are the prices which determine the numbers of people and mass of cargo arriving and the trip frequency. Among the constraints are various supply chain challenges including propellant production capacity and available payload capability per launch system, including maximum number of passenger seats per launch system.

How Does It All Fit Together in a Value Chain? Market sector 1 has interdependencies with all the other market sectors. In the Early Phase, transportation providers buy propellant and use spaceports and order systems from prime contractors (which in turn are supported by suppliers and subcontractors). Spaceports hire construction firms and support equipment, and they use cargo handlers to prepare payloads. They can become a major generator of employment and other economic benefits to a region or a country. Again, in the Early Phase, propellant providers are assumed to be Earth-based operations, and all customers and providers are assumed to be Earth-based. In the Mature Phase, however, it is assumed that some propellant providers will be Moon-based.

A summary of this market sector is provided in Fig. 7.1. We adopt this same format for each of the 11 subsequent market sectors, to make it easy to make comparisons.

How Can the Data Be Improved? Simply by continuing to monitor and update, as new specs and pricing is published.

OK, so how do you feel about investing in this sector? You have a choice of many possible providers, some already well established. And their supply chain contributors are also generally well known and understood. So, you have alternative entry points within the value chain to consider. This is certainly the least-risky sector for investment today—as reflected in the lowest range of uncertainties. Some of the future lunar commerce segments may offer bigger rewards, but at a higher level of risk. We'll figure out the revenue potential later in this primer (Chap. 11), but for now, this represents the most visible face of the return to the Moon, and to its future development. How reliable will the former GLXP team contributors turn out to be? NASA has taken a certain risk in using this nontraditional way of obtaining services.

References

Adventurers' Survey. (2006). Webber, D and Reifert, J. www.SpaceportAssociates.com

CLPS. (2018). nasa.gov/commercial-lunar-payload-services-overview

GLXP. (2015). *The Google Lunar XPRIZE—Past, present, and future.* RiSPACE Oxford. Barton, Webber and Wong.

LCP. (2022). *Annex A and Annex B Excel model data.* Moonvillageassociation.org/download/the-lunar-commerce-portfolio-first-edition-november-2022

Lunar Gateway. (2023). en.wikipedia.org/wiki/Lunar_Gateway

Orion. (2020). *Lockheed Martin spacecraft.* NASA.gov/topics/moon-to-mars/getting-there

SLS. (2020). NASA.gov/topics/moon-to-mars/getting-there

8

Moving About

This market sector, Transportation on the Moon, is #2 in the Lunar Commerce Portfolio classification scheme, described in Fig. 6.1. It was considered as only applying to lunar surface activities. And for our purpose, we also included all EVA (i.e., extra vehicular activities) systems including space suits. You might want to change that, if you are doing your own forecast modeling, just so long as you make sure to include it somewhere else. The main categories of rovers or hoppers were assumed to be either robotic or crewed craft. And within the crewed category, we separately considered pressurized and unpressurized rovers. The precursor craft during the Apollo program was the Lunar Roving Vehicle (LRV) (LRV, 1971), which could carry two astronauts wearing spacesuits. This sector 2 is also relatively well understood at this stage, compared to the subsequent sectors 3 thru 11. You probably already know some of the players. Hoppers, by the way, have not so far (Aug 2023) been deployed on the Moon—although a helicopter is operating on Mars. Hoppers provide the mobility of helicopters, including the aerial perspective, on a planetary surface without any atmospheric pressure. A craft lands, then takes off again, and translates sideways, before landing a second time (and potentially many more times).

Within the Artemis program, the initial focus has been on the small robotic rovers which have been procured as services via the CLPS program discussed in Sect. 7. Several contracts have already been signed for these precursor services. As mentioned above, several of them were initially designed as part of the Google Lunar XPRIZE competition, and now form the baseline for the Artemis robotic systems. GLXP judges observed tests of both the Team Indus, and Team Hakuto, rovers in a lunar simulated environment in October 2017 in Bengaluru, India. The judging panel was observing such things as the

D. Webber, *Lunar Commerce*, https://doi.org/10.1007/978-3-031-53421-8_8

slippage of the rover wheels as they progressed over the lunar regolith simulant, and trying to calculate how this impacted the measurements of distance traveled, because of course there is no GPS on the Moon, and we had to be able to measure distance traveled in order to award the relevant multimillion-dollar prizes for lunar performance. Years later, as this book is being written in August 2023, the Indian lander Vikram of the Chandrayaan-3 mission has landed on the lunar surface, and deployed its rover Pragyan, making India the fourth country (after the USA, Russia, and China) to soft-land on the Moon. One cannot help but notice the similarity of the design with that of the Team Indus GLXP craft. Furthermore, a new generation of rover is being designed and procured as part of the US governmental NASA human surface mobility provisions, seen as the crewed successor to the Apollo-era LRV. Initial design and demonstration contracts are being awarded.

Regarding spacesuit design, there was considerable feedback from the 12 Apollo astronauts who walked on the Moon to the extent that there would need to be improvements in design for longer duration missions. In particular, they drew attention to the problems of lunar dust. The regolith's abrasive and electrostatic properties, it turned out, made it very difficult to keep the spacesuit linkages and screw fittings operational. Moreover, the dust which accumulated on the suits affected temperature control. Also, special attention was needed with regard to the gloves, because of the pressure differential, which resulted in extremely stressful effort needed to perform even simple operations using the gloves. Tom Stafford, Commander of Apollo 10, gave testimony in Washington DC in February 2004 to the Aldrich Commission, where I noted his remembered frustration when he declared "It's like doing a penmanship contest wearing boxing gloves."

As the potential period of lunar commerce approaches, there will be a need for a whole new era of robotic and crewed vehicles on the lunar surface, some working as runabouts, and some engaged in what on Earth would be labeled as "Earth-moving" activities. They would include haulers, diggers, bulldozers and drillers. We took particular care in the work of the Lunar Commerce Portfolio to avoid double-counting of these vehicle classes, so it was clear what would be included in sector 2, and what might be in, e.g., sector 5 or sector 7.

And we have already mentioned, under our assumptions, lunar surface space tourism may be a significant driver of the commercial lunar economy. Lunar space tourists will be wanting to explore the surface of the Moon around the lunar base, and maybe in particular to visit historical and/or photogenic sites on the surface (as shown in Fig. 6.2). There are over seventy such sites, some with more historical consequence than others. So, we have included the anticipated demand for robotic or crewed "tour buses" within this category also (Fig. 8.2). Where do *you* imagine that you would want to go during your stay on the Moon? Wouldn't it be neat to go visit the tracks made by the first lunar rover, the Russian Lunokhod (Lunokhod, 1970)? What about trying to find where Alan Shepard's golf ball landed, after he hit it with a lash-up golf club at the end of the Apollo 14 mission? At the time, he said it went for miles, but his former Project Mercury buddies had their doubts (Shepard Golf Shot, 2021). As mentioned in Chap. 7, the GLXP competition had a separate prize category for taking high-definition pictures of lunar heritage sites, but also had rules to ensure that the sites suffered no damage from such activities. Subsequent to the end of the GLXP competition (March, 2018), the organization with the weird name ForAllMoonkind (with Observer status at the UNCOPUOS in Vienna) (ForAllMoonkind, 2017) was created to continue to protect those lunar legacy sites—which include such historic artifacts as Neil Armstrong's first bootprints on the Moon. It is expected that those bootprints still remain, in the absence of wind or rain on the Moon, but there is radiation, and there are meteoritic bombardments, and even Moonquakes, so it would be of scientific interest to be able to continue to observe them over time to see what effects these factors might have.

We now consider the systematic key items which we have assembled with regard to this market segment as reflected in the findings of the Lunar Commerce Portfolio, Version1.

What Is the Formal Description of the Segment? The market sector consists of three distinct categories, none of which takes place in lunar orbit:

1. Robotic rovers or hoppers provided or leased by commercial robotics firms conducting ongoing surveying/resource mapping/or diggers and haulers supporting the mining businesses
2. Crewed rovers or hoppers provided or leased by commercial taxi firms, including providing site visits for lunar tourists. Rovers may be either pressurized or unpressurized.
3. EVA systems provided or leased hardware, including spacesuits to service the need to leave the fixed habitats.

At present, there are no universally accepted standards for either rover payload hosting, or for spacesuit design. However, it will clearly be advantageous as the era of lunar commerce proceeds that commonly accepted (i.e., internationally accepted) standards be adopted—at least in so far as interfaces and communications links are concerned.

During the Early Phase, the transport on the Moon will support surface survey, scientific experimentation, infrastructure construction, and maintenance. It is assumed in that period there will be no lunar surface tourism activities. During the Mature Phase, there will be additional activities, such as those connected with lunar tourism, entertainment and leisure, as well as with lunar manufacturing and mining. In fact, there will be cross-interactions with all of the other market sectors.

Who Are the Potential Suppliers? We need to look at robotic transportation, crewed rovers—both pressurized and unpressurized, and EVA systems (spacesuits). There are a great many potential suppliers of the robotic surface transportation rovers, including Astrobotic, Blue Origin, Ceres, Deep Space Systems, Draper, Firefly, Intuitive Machines, Lockheed Martin, Maasten, Moon Express, Orbit Beyond, Sierra Nevada, SpaceX, Tyvac, and General Motors. Three GLXP teams had previously been able to win interim-risk reduction milestone prizes in the "mobility" category (Astrobotic of USA, Hakuto of Japan, and Part Time Scientists (PTS) of Germany, each won $0.5 M). It was very encouraging as a GLXP judge to watch the young competitors testing their rovers in "dirty thermal vacuum" conditions, and furthermore conducting simulations via their personal PCs. This new generation of engineers worldwide is the hope of the future—and they don't appear to need much in the way of supporting infrastructure to achieve impressive results. There are also possible rovers from China (China Moon Rover, 2013). None, at present, are considering providing hoppers, although in the GLXP competition, Astrobotic did demonstrate some preliminary capabilities in this area, before judging panel members at a commercially available test facility at Kennedy Space Center in Florida. They were attempting to win a GLXP interim "mobility" prize, but could not convince the judges that they were able to control their prototype (which was swinging wildly about, suspended from a crane, and not "translating" horizontally, as it was supposed to do). In considering the crewed rovers, most offerings are unpressurized (including Astrolab's Flex Rover). It is probable that the pressurized rover category will not be available in the Early Phase, but offerings are being developed by Toyota for JAXA use, and within NASA, for use during the Mature Phase.

With regard to the spacesuit design and service, both NASA and SpaceX are likely to have their own systems. Providers of lunar tourism operators may choose to purchase their own variants, rather than renting from a provider. Within the Lunar Commerce Portfolio Version 1, information was assembled on three commercial spacesuit firms, and 11 potential providers of robotic and crewed rovers. Within LSIC, studies have been undertaken, which conclude that for sustainable lunar occupancy, airlocks would need to be mandatory, and that lightweight suit covers should be developed for the early Artemis short duration lunar stays and EVAs. NASA is pursuing a Dust Solution Testing Initiative (DuSTI) and among the approaches being tested are surface coatings (polycarbonate and fused silica) for both hard and soft goods, such as spacesuits. Also, in this initiative (described at an LSIC meeting in May 2021) would be the development of pliable gel/putty/clays which could be used to remove lunar dust that had been brought into a habitat environment as a result of EVA activities.

Examples, from among the six potential suppliers of robotic surface transportation, are the Astrobotic and Lunar Outpost offerings. Astrobotic's offering has its origins in the firm's attempt to win the Google Lunar XPRIZE, where they succeeded in winning some intermediate awards. They have now been the recipient of a NASA CLPS service contract, and they aim to deliver the VIPER rover to investigate lunar volatiles, particularly water, on a mission currently scheduled for 2024. Lunar Outpost, in contrast, was not one of the Google Lunar XPRIZE spacecraft contenders. The firm is developing the Mobile Autonomous Prospecting Platform (MAPP), which is contracted to be delivered to the Moon onboard an Intuitive Machines lander in 2023/2024. The LCP provided details on five potential suppliers of crewed rovers of various types. It is not expected that any of them will be operating during the Early Phase, however. Two examples will suffice to illustrate the possibilities, and we will describe the Lockheed Martin and Toyota crewed surface transportation vehicles. Lockheed Martin is working with General Motors and MDA of Canada to produce a crewed or robotic Lunar Mobility Vehicle (LMV) which is a contender for a NASA development and service award for a Lunar Terrain Vehicle, due to be chosen following a June 2023 RFP process. The MDA part of the vehicle will be a mobile robotic arm. The Toyota offering is a pressurized crewed rover, being developed with JAXA. It is intended to use hydrogen fuel cells as a power source for missions of up to six weeks' duration. LCP Version 1 did not specifically identify any potential providers of those robotic vehicles (diggers, haulers, bulldozers, etc.) needed to support the lunar mining sector (sector 7).

Who Are the Potential Customers? There is quite a bit of overlap across the customers for each of the three identified parts of this market sector. They would include space agencies, NGOs, academic and research institutions, and the customers represented by the other identified market segments, such as those involved in lunar construction, manufacturing and resource extraction. Even the lunar agriculture producers may use transportation services to deliver centrally produced foodstuffs to customers. There is likely to be a significant lunar tourism engagement once we are in the Mature Phase, with pressurized surface vehicles being used to take tourists to view legacy sites.

What Are the Likely Drivers and Constraints? Major drivers will differ, depending on the kind of rover or hopper being considered. It is unclear, at the time of this version 1 of the LCP, how pricing will be established—particularly for lunar surface tourists wishing to travel across the landscape to heritage sites. Will the price be bundled in with the assumed very high ticket-price for the lunar vacation? On the constraint side, there may be limitations in transportation, pending the full establishment of the landing pads and sintered roadways.

How Does It All Fit Together in a Value Chain? This will vary somewhat between the three subsectors of robotic rovers, crewed rovers (both pressurized and unpressurized), and EVA systems. The robotic rover class have already begun to be deployed, with several of them having origins in the Google Lunar XPRIZE competition. These Early Phase robotic rovers are very limited in capability, but nevertheless will be performing important preliminary experiments prior to human landings. Also, during the Mature Phase, there will be entirely new categories of robotic rover, which will be deployed as part of the equipment involved in mining and manufacturing operations on the Moon. The crewed rovers will be deployed partly to support governmental and contractor astronauts on the lunar surface, as well as for conveying future lunar surface tourists. The suppliers of end-user lunar surface transportation services may exhibit various degrees of vertical integration along the value chain. The providers of those Early Phase robotic payloads, which are designed for exploration activities, will probably subcontract the requirement for sensors. While crewed rovers will almost certainly be capable of autonomous and teleoperated mobility, it is likely that any vehicle intended for use by lunar tourists will also require an operator/host for the duration of surface trips. In the case of spacesuits, and other elements of the EVA market sector, it will be very important to have an effective maintenance and repair operation, and the ability to provide full support regarding fitting and communications. It is assumed that for a considerable time, into the Mature Phase, spacesuits will still be manufactured, and delivered, from Earth. Thereafter, the steady state operation on the Moon will be maintained via the careful repair and servicing of suits which were delivered earlier. The original manufacture on Earth of such equipment involves a very long list of component manufacturers, who in turn are reliant on a wide range of raw materials providers, all of which are fully documented in the Lunar Commerce Portfolio. There will of course be a great deal of integration with the mining and manufacturing sectors, in provision of robotic rovers with specialized functions, such as diggers, regolith movers, sintering equipment, etc., but all sectors will be using the lunar

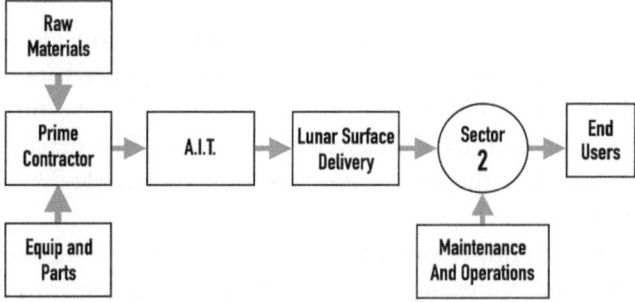

Market Sector #2 Transport On The Moon

Description   Robotic Rovers/ Hoppers, Including Haulers, Diggers, Bulldozers. Crewed Rovers, Hoppers, Including Tourism Vehicles – both Pressurized and Unpressurized, EVA Systems and Spacesuits

Potential Customers   Space Agencies, Academic And Research Institutions, Lunar Tourism Operators, Other Market Sectors (5, 7, 8, 9, 10).

Potential Suppliers   Astrobotic, Ceres, Draper, Blue Origin, Deep Space Systems, Firefly, Intuitive Machines, Lockheed Martin, Masten, Moon Express, Orbit Beyond, SpaceX, Sierra Nevada, Tyvak, General Motors, Astrolab, JAXA, Toyota, etc.

Drivers And Constraints
Availability of Roads, Capacities, Pricing, Development of Tourism Markets, Regulations for Heritage Sites, etc.

Value Chain Schematic

Fig. 8.1   Summary of LCP data for sector 2—Transport on the Moon. (Credit: DW/MVA)

surface mobility equipment at various stages of their respective operations. So, what do you think? Does this look like a good place to invest?

A much-simplified summary of this market as recorded in full in the LCP is provided in Fig. 8.1:

How Can We Improve the Data for This Sector?   Some recent NASA procurement activity is going to result in a better understanding of lunar surface options, which will help. If you work with a UN-related regulatory agency, you may want to reflect on what is being proposed for this new kind of rover on the Moon. And, by the way, consider if you want to preserve any of the

Fig. 8.2 An example of a class of potential crew-carrying pressurized lunar rover vehicles. (Credit: JAXA/Toyota)

original rover tracks, or footprints, from the Apollo era, before the new generation begins operating.

You might think that this is a good market sector for considering an investment. It is new, and there are several new prospective entrepreneurial startup providers, where you could get on board with an early investment opportunity. Moreover, the technology is well understood. But maybe you should wait until we see what the likely revenues are that emerge from the model, later in this primer.

Figure 8.2 gives an impression of an early prototype pressurized lunar rover, which could be used, for example, to take lunar tourists on trips to visit lunar heritage sites, such as are demonstrated in Fig. 6.2.

References

China Moon Rover. (2013). http://en.wikipedia.org/wik/YUTU_(rover)
ForAllMoonkind. (2017). https://forallmoonkind.org
LRV. (1971). *Lunar roving vehicle (moon buggy).* https://en.wikipedia.org/wiki/Lunar_roving_vehicle
Lunokhod. (1970). https://en.wikipedia.org/wiki.Lunokhod_1
Shepard Golf Shot. (2021). Smithsonianmag.com/Smithsonian-institution/when-astronaut-alan-shepard-hit-golf-shot-heard-round-the-world-180976903/

9

Infrastructure

So, we know how to get there. And how to move about when we have done so. But that just about establishes a beach-head, rather than a station, or space-port, on the Moon. We have already noted how harsh an environment we face on the Moon. We could not be expected to survive very long in such a place without a few other basics in place. It might take a few trips before they can all be established. And it would be an unpromising way to start trying to operate a business before some of these essential infrastructure elements would be in place. None of them are of course there today. We could simply continue to live in our arrival vehicle, just as was done with Apollo, but that would not lead to a permanent settlement, so we need more. The initial landings under the US-led Artemis program are designed to be at the lunar poles, because that is the most likely place where water might be found. Due to astronomical geometry, there will be certain zones there, in relatively close mutual proximity, which will either be in permanent darkness, or in permanent sunlight. So, that means a permanent energy source. Under this chapter heading, we consider market sectors 3 thru 5, which represent an essential prerequisite infrastructure needing to be established, before we can contemplate the full blossoming of a lunar economy reflected in sectors 6 thru 11. The two organizations mentioned earlier, LSIC and LEAG, have been steadily working in these areas, sometimes including Apollo veterans within their group of expert advisors. I have taken part in Zoom calls during the last year when both David Scott and Harrison Schmitt (who walked on the Moon in 1971 and 1972, respectively) were able to offer practical advice for the new generation of lunar explorers, 50 years on from their own ventures (Schmitt, 2021; Scott, 2021). And, as recently as August 2023, the DARPA agency has also latterly shown

D. Webber, *Lunar Commerce*, https://doi.org/10.1007/978-3-031-53421-8_9

interest in contributing to this necessary infrastructure and has begun to solicit proposals to address elements of these needs (DARPA, 2023).

What do you think you would need to get beyond that initial "flags and footprints" landing vehicle? I suspect you would need quite a lot, assumed to be put there by the government or governments, before we could try to make a commercial living on the Moon. These questions were faced by the analysts putting together the Lunar Commerce Portfolio. They concluded that of course we would need to know we had a supply of water, oxygen and food for starters. Later on, the plan would be for these supplies to be obtained locally by commercial means, but right now, we would want them to be delivered up front. Things have changed, of course, since humans arrived at previously unknown parts of the Earth, and decided to live there. Nowadays, in those circumstances, we would also want to have available a source (or sources) of electrical power. We would want to be able to communicate between both fixed and moving groups of fellow-humans. Ideally, we would want to know, and have available, a means of navigation and position location (even though, in principle, and *in extremis*, this could be done by reference to the stars in the heavens).

As in the olden days, when we would have needed a port for our ships, we know that in this era, in this place, we shall need landing pads. Quite apart from the problems that lunar dust creates for crews who are landing their craft on the surface, regular landings and take-offs on the Moon will cause a very difficult operating environment for those inhabitants who are already there, unless landing pads are created with berms to protect surrounding infrastructure and habitations. This was clearly not a problem that Neil and Buzz had to face. Neil noticed the peculiar behavior of the lunar dust during landing. It shoots out horizontally at great speed from under the lander during descent, and then ceases immediately at engine cut-off. All the particles of regolith flew in shallow parabolic trajectories, unimpeded by any trace of atmosphere, and in a second or so returned to the surface. It just disappeared in an instant. No dust remained in suspension. Weird. Landing in a flying sand-blaster is certainly not a good way to introduce yourself to new neighbors. The same argument applies to roadways. The lunar regolith is an unforgiving medium which will make life very difficult for settlers, unless protections are put in place—and this would involve establishing roadways between parts of the base which will be subject to the most wear. Remember, too, that we are effectively building a way-station (due to the Moon's favorable gravity well), to the rest of the solar system. It will be Spaceport Moon, and would be the point of departure, and return, and of refueling, for future interplanetary missions going to Mars and points beyond. The lunar outpost will need to handle frequent arrivals and departures. We shall learn later, in discussing regulations, that noninterference is an important concept in international law. We'll need those berms.

For the purposes of the Lunar Commerce Portfolio, we considered these infrastructure elements under the separate sector headings of Sector 3 Communications and Navigation, Sector 4 Energy and Power, and Sector 5 Civil Engineering and Construction of Landing Pads and Roadways.

The role of government, apart from putting these infrastructure pieces in place, would be to identify the need and establish common interfaces—ideally internationally. They will be continuing even at this early phase to push the boundaries of science and exploration, to carry out basic ISRU research looking for water and oxygen, and even exploring different zones of the Moon, including on the Far Side, as part of their scientific and discovery needs. For instance, a fairly high priority scientific objective would be to establish a farside radio telescope—a venture with very little obvious commercial market attraction, and so must be funded by government. Governments will probably also be interested in deep drilling, which again initially would not attract commercial investment funding. This Infrastructure establishment is another area where the particular future needs of the lunar space tourism community will need to be considered. What will the billionaire tourists be expected to demand as a minimum, even as they regard themselves rightly as pioneers, before they would consider coming to stay (and pay for) an experience of living on the Moon. We shall discuss the longer-term needs in Chap. 10, but we need here to at least establish the minima. Billionaires can be rather fussy about their basic needs.

We now treat these three named sectors in turn, giving each of them the full systematic treatment with which you have become accustomed.

Market Sector 3—Communications and Navigation

What Is the Formal Description of the Segment? Within this sector is included both the navigation needs for maneuvering on the lunar surface, and the communications needs (including IT, data exchange, detection services, and Internet) of lunar inhabitants and operators. They are separately identified for operating in the lunar-surface-to-lunar-surface, lunar-surface-to-orbit, and lunar-surface-to-Earth domains. Lunar data is expected to be a significant sector. The needs may be satisfied by either governmental or commercial operators. The Early Phase will see the majority of demand coming from institutional stakeholders, several of whom operate within the interagency operations advisory group (IOAG) (IOAG, 2000), with potential products and services involving the Lunar Pathfinder, LunaNet, and ESA's Project Moonlight. During the Mature Phase, the demand would increase, and additional noninstitutional users related to the lunar surface tourism businesses would be added. Folks made a lot of money on Earth in these nav/com market sectors—what about the Moon for a repeat? The trouble is, we're starting from scratch on the Moon, and it could be an expensive outlay at the start. And we need to know if there will be enough customers to support the necessary investments. So, this is maybe a job for governments. What is peculiar about the Moon when it comes to communications and navigation? First of all, there is no ionosphere. And so, all signals will have to operate via line-of-sight. And at the poles, and in mountainous regions, signal repeaters will be needed due to blockage by terrain, including crater rims. There is no equivalent to the Earth's geostationary orbit, excepting maybe the Lagrangian Points. So, an eventual lunar comsat constellation will probably require tracking antennas at the surface to maintain the communications links.

Satellite navigation services will also be needed to provide lunar inhabitants with the necessary guidance as they move about the surface. You cannot, of course, use a magnetic compass on the Moon. Accurate navigation and position location will be needed for a range of situations, not least of which is the ability to know enough about location on the surface to be able to avoid interference with other lunar operators.

Who Are the Potential Suppliers? Space-based nav/com systems on Earth require, in general, three different and distinct kinds of manufacturer and supplier, and for the Moon, the same will apply. There will be the space segment, i.e., the communications satellites, and the means to get those satellites into the correct lunar orbit. Then there will be the manufacturers of the major receiving station antennas on the surface. This on Earth is referred to as the Ground Segment—and has its own specialist manufacturers. And then, there will be the providers of the end-user equipment, such as hand-sets. Again, on Earth, these products are made by a different specialist group of manufacturers. And the Moon will be no different. There could also be a nonspace-based solution, using cellular-type technology, and that would probably involve yet another kind of supplier.

Within the LCP database, we note the following potential suppliers. One is Kepler Communications, with its proposed AETHER network, which involves a Ku-band, S-band, and optical data service. Individual component manufacturers could include Alen Space, Tethers Unlimited, Nokia, and Honeywell Aerospace. Advanced Space could provide the navigation systems. Within the database of the Lunar Commerce Portfolio, Version 1, there were

78 identified potential commercial suppliers for this segment. Eighteen of these were providers and operators of communications and navigation satellite systems used for terrestrial use. Among the various potential operators were 10 providers of terrestrial GPS services and 8 handset manufacturers. Then there were 7 operators of the huge terrestrial Earth stations needed for providing mobile users with communications or navigation services, and a further 35 assorted manufacturers of equipment and subsystems for terrestrial nav/com services.

We can draw a few examples from this database to give you an idea of what might be needed, and what is possible. Currently the dominant method of providing communication and navigation services at lunar distances has been through tracking stations such as NASA's Deep Space Network, or ESA's Estrack. NASA's Deep Space Network is operated by NASA's Jet Propulsion Laboratory (JPL) with facilities in Goldstone (USA), Madrid (Spain), and Canberra (Australia). It has been used since the sixties but by now has limitations and challenges for the modern era. Around Mars, for example, it was necessary to add infrastructure called the Mars Relay Network of five spacecraft in orbit to support exploration activities at the Red Planet. ESA's Estrack is similar, having a system of seven major ground stations (in French Guiana, Belgium, Portugal, Sweden, Australia, Spain, and Argentina) and a control center in Germany. The international Interagency Operations Advisory Group made a recommendation for four lunar relay orbiters to be deployed initially in lunar orbit to support early missions, and for some coming-together of the various operational standards. The Group suggested that the following architecture should be developed and deployed:

- lunar space internet
- lunar relay network
- lunar surface network; and
- a high-speed trunk link that connects back to Earth.

Some first examples of governmental responses to these needs are NASA's Lunar Pathfinder, ESA's Project Moonlight, and ESA's Lunanet.

Who Are the Potential Customers? Among the established space agencies included in the LCP database are NASA, ESA, JAXA, CNSA, Roscosmos, and CSA. Newer agencies becoming involved include LSA (Luxembourg), UAESA (United Arab Emirates), PhilSA, NZSA, and ASA. Universities and research institutes will be interested, as will the many private companies operating across the gamut of the other market sectors included in the portfolio.

What Are the Likely Drivers and Constraints? Telecommunications and navigation have proven to be major businesses on Earth based upon space infrastructures. And there is some expectation that this could also happen on the Moon. However, there is at present very little clarity about what will be provided by government, and what may be expected by commercial providers, and this certainly applies to matters of pricing, which could be a major lever of driving demand, and which may need to be very high in order to recover development costs. Regarding constraints, there is at present no equivalent of GPS on the Moon, and until the appropriate space segment is operational, this will limit the development of related services and of lunar handset devices. Because of the likely relatively low numbers of users of the system, when compared with terrestrial operators, there may be a disincentive to become too involved at the outset.

How Does It All Fit Together in a Value Chain? There is a generalized value chain which provides an indication of supplier dependencies, and is equally valid for both the Early Phase and the Mature Phase. But at this stage, it is difficult to include much more detail, due to the uncertainties. Essentially, both for the Early Phase and the Mature Phase, it will be necessary to manufacture equipment on Earth and then have it transported to the lunar surface. Of course, there would be different detailed value chains for each of the three kinds of product and service provider we have described.

How Can the Data Be Improved for This Sector? Basically, we must await some clarification from the various national space agencies regarding how they intend to work together on this. Together with initial thinking from international regulatory bodies, such as the ITU and the UN's COPUOS, that would lead to some better understanding of forecast possibilities and limitations. It's already tough enough on Earth for all of us to stay up to date on the latest communications gizmo device and operating system. We are going to need some agreed standards for the Moon.

A basic summary of this market sector is provided in Fig. 9.1:

Market Sector 4—Energy and Power

This is clearly a key part of the lunar infrastructure. We have to have energy to be able to do *anything*. Probably, the main sources will be solar and nuclear. As with any energy companies on Earth, it will be necessary to even-out demand throughout the lunar day (including a two-Earth-week long lunar

Market Sector #3 Comms And Nav

Description   Provision of Communication and Navigation Services for Lunar Surface and Lunar Orbit Fixed and Mobile Customers, Including Voice, Data, IT/Internet. Orbit to Earth, Surface to Orbit, Surface to Surface.

Potential Customers   Space Agencies, Universities, and Research Institutions, Commercial Operators – Including Tourism Companies. All 10 Other Market Sectors.

Potential Suppliers
Lockheed/Crescent, Kepler, General Dynamics, Nokia, Honeywell, Tethers, etc.

Drivers And Constraints
International User Agreements, Pricing, Capacity, etc.

Value Chain Schematic

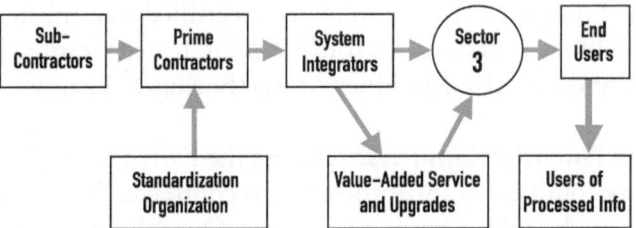

Fig. 9.1   Summary of LCP data for sector 3—Communications and Navigation. (Credit DW/MVA)

night)—and this will require energy storage systems like batteries. We now go to our standard format to describe what is contained in the LCP.

What Is the Formal Description of the Segment? The main elements of this market sector are power generation, power distribution and power storage. Most of the power generation in the Early Phase will be through solar panels (both fixed and mobile), radioisotope thermoelectric generators (RTGs) and hydrogen fuel cells. The storage will be via batteries. In the Mature Phase, there will be an additional power transmission and distribution network (involving transformers, cabling and circuit breakers). The extra power demand may be satisfied by chargeable atomic batteries (CABs), and small modular reactors (SMRs). Also, it is likely that solar cells will be able to be manufactured *in situ* on the Moon out of the silicon-rich regolith, and that this process will be largely robotic. We shall need to take account of the astrophysics of the Moon's axis of rotation, which provides for a few isolated peaks of eternal sunlight at the poles. Solar arrays in these polar regions may ultimately be the source for most of the power on the Moon, because locations elsewhere suffer from the disadvantage that they cannot generate power for two Earth-weeks at a time. This will underline the importance of high-conductivity cabling, or microwave transmission facilities, for moving power about across the lunar surface, as well as the need for energy storage systems. In this market segment, for the LCP, we consider only lunar surface demand.

Did you ever wish you had shares in an energy company on Earth? Well, what about the Moon—starting all over again? This is a sector where it is unavoidable to consider the need for building a complete centralized energy support system for the Moon, implying either a unified council, or brokerages, to pave the way for the use of power for all purposes on the Moon. For the Early Phase, it is assumed that the various missions will bring their own power sources, and that therefore there will be no market for power transmission, although as part of a NASA contract, three consortia have won awards for conceptual design work for a fully flight-certified transmission power system for the Moon: Lockheed Martin/BWXT/Creare, Westinghouse/Aerojet Rocketdyne, and Intuitive Machines/Maxar/Boeing/X-Energy.

Who Are the Potential Suppliers? During the Early Phase, potential suppliers will be any company or institution that is contracted by a national government agency, such as those mentioned above. Once the energy framework for the Moon is established, i.e., during the Mature Phase, countries will likely provide a consortium approach, similar to how energy suppliers meet demand on Earth. During the Mature Phase, to address the three stages of generation, transmission and storage, we can add to the above Early Phase list: Airbus, AGPower92, Ultra Safe Nuclear, Maana Electric, Solaren Shimizu, Xeno, Regher Solar, Ion Power, Howe and Longi, X-Energy, ETP, Photonicity, Instarz, ASDA Technologies, Eagle Picher, and Infinity. As examples of energy generation supply companies, we can cite AGPower92, for instance, which is proposing a small utility power source modular reactor (5–10 Mwe), and Shimizu Corporation, which plans to build a lunar power plant using locally sourced regolith. For distribution operating companies, Extra Terrestrial Power (ETP) is focusing on power transfer options, and for Energy Storage, we can reference Eagle Picher, which is a battery manufacturer who has already seen products flown in space. Lockheed Martin is designing architectures, such as their Lunar Vertical Solar Array Technology (LVSAT), and presenting them to LSIC meetings (e.g., May 2021 Spring Meeting). In the same forum, other potential providers were using modeling simulations to demonstrate the optimum height for tall towers near the lunar poles for taking advantage of the Peaks of Eternal Light.

Who Are the Potential Customers? In the Early Phase, most missions are expected to be providing their own power sources. In the Mature Phase, a different process is envisaged. Several space agencies (e.g., NASA, ESA, JAXA, CNSA, Roscosmos, CSA, ISRO), government organizations (e.g., the DOE of the USG) and commercial companies (SpaceX, BWXT, Blue Origin, General Atomics, Lockheed Martin, Astrobotic, Eternal Light, Axiom, and iSpace) will be the end customers for power systems provided by utility providers. A major slice of power will be needed by the mining and manufacturing industries. In a survey conducted within the LSIC organization, and reported in March 2022, it was stated that 100 kw would be needed to produce 100 tons of H_2O, and 430 kw would enable 50,000 tons of regolith to be excavated.

What Are the Likely Drivers and Constraints? We are still at a very early stage of defining how power will be generated and distributed on the Moon, whether from solar or nuclear sources. Some potential lunar manufacturing and mining operations will need to use a lot of power, and since power will at least initially be a scarce resource, it will need to come at a high price. Will battery technology keep up with the expected demand, especially for living and working thru the lunar night? We await the results of the first explorations of solar power generation from the "peaks of eternal light" at the poles, and the development of the cabling or microwave surface distribution networks to understand how this sector will develop. Within the LSIC framework, various architectures have been proposed. Ongoing work within the LSIC framework is addressing various parts of the architecture needs. USNC-Tech is developing chargeable atomic batteries, for instance, and Astrobotic is working on an integrated lunar surface power grid, and SWR Technology is developing an electrical interface which will be resilient on the Moon. Needless to say, all of these ideas will need to be tested in a lunar environment in order to reduce the number of unknowns that we are facing at this stage.

How Does It All Fit Together in a Value Chain? It recognizes the three separate task areas, of power generation, power transmission and power storage, which feed into a "Utilities" overview function. For all three areas, there will be a need to make equipment on Earth, and then have it delivered to the lunar surface. But it is unclear how to refine the value chain much further at this time, due to the uncertainties in how the national space agencies are intending to manage and operate this market segment.

How Can the Data Be Improved? Again, we need to ask how we can get a better handle on the data assumptions we are using. As with the other sectors covered in this chapter, we are addressing essential infrastructure. And this infrastructure will only be put in place when governments have come to some common understandings about common interfaces, etc. And, even more fundamentally, why we are collectively doing this? Once this has been done, we shall have less uncertainty in the forecast outcomes. A summary of this market sector is provided in Fig. 9.2:

Market Sector 5—Civil Engineering-Construction of Landing Pads and Roadways

We explained earlier in this chapter why this sector is so important to lunar infrastructure. It has already been the subject of much experimentation, specifically with respect to figuring out how to perform sintering in a vacuum using various lunar regolith simulants. We now use our standard format to capture a summary of the relevant sections of the LCP.

What Is the Formal Description of the Segment? This sector undertakes the requirements of operations and maintenance of launchpads and roadways. In the case of both the launch and landing zones, and the roadways, sintered regolith techniques are expected to be used, and the zones will be surrounded by berm protection surrounds. In the Early Phase, there is likely to be very little (mainly experimental) and only *ad-hoc* arrangements, while in the Mature Phase, the zones may be built, operated and maintained as a regulated Moon Port infrastructure, once data has been collected about damage and deterioration rates. By definition, we are only interested for this market sector in lunar surface (not orbital) activities. Know any successful civil engineering contractors on Earth? There are more than a few. Can they repeat the success in a new ballpark—the Moon? Or maybe they will leave the field to new

Market Sector #4 Energy And Power

Description   Provision of Power to Fixed and Mobile Operators on the Lunar Surface by Solar or Nuclear Means. Generation, Distribution & Storage

Potential Customers   Space Agencies, Universities, and Research Institutions, Commercial Operators Including Tourism Companies. All 10 Other Market Sectors.

Potential Suppliers Lockheed Martin, Airbus, Aerojet Rocketdyne, General Atomics, Maxar Technologies, BWXT, USNC, Maana Electric, ADA Technologies, AGPower, Regher, Infinity, etc.

Drivers And Constraints
International User Agreements, Pricing, Capacity, Interfaces, etc.

Value Chain Schematic

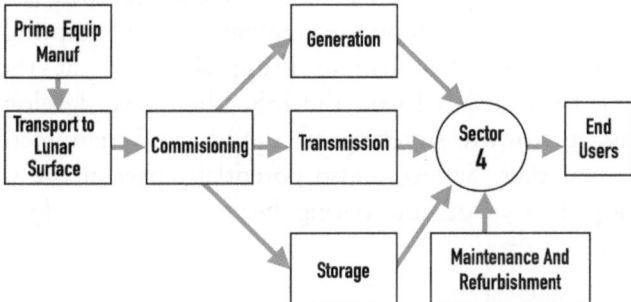

Fig. 9.2   Summary of LCP data for sector 4—Energy and Power. (Credit: DW/MVA)

entrepreneurial lunar operators? At an October 2020 LSIC meeting, a NASA technologist (Rob Mueller) underlined the potential difficulties of the task. He stated that the lunar landing/launch pads (LLP) would need to be able to withstand gas temperatures of 3000–4000 °C, and gas velocities of 2000–3000 m/s. The LSIC Working Group that is addressing the issue and mitigation of lunar dust has noted the negative effects in the following areas: optical systems, thermal surfaces, fabrics, mechanisms, and seals. This is a Big Deal. Particularly when considering a sustainable human presence on the Moon. An LSIC Dust Mitigation Workshop in February 2021 was attended by over 300 attendees (43% industry, 28% government, 20% academia, and 9% nonprofit), and identified the following technology gaps needing work: abrasion resistant, passive dust repelling optical surface coatings, dust tolerant, abrasion resistant dynamic seals, and dust monitoring/filtration equipment for inside habitats. Terrestrial analog work on sintering and alternative approaches (bricks and tiles) for lunar landing pads has been carried out since 2008 on Hawaii's volcanic terrain, and is regularly reported in the LSIC meetings. Some of the field tests even used GLXP team rovers to help (e.g., Team Puli in 2013). The Apollo astronauts used Hawaii to conduct training exercises back in the sixties (as well as Iceland, another volcanic terrain). However, it seems that even with all this prior work, there is still no current preferred approach to building the landing pads and roadways.

Who Are the Potential Suppliers? One can anticipate various A&E firms, as well as specialized equipment providers, and various miscellaneous hardware and services suppliers. Bechtel, for example has demonstrated a strong interest (and they were the original builders of the LC40 launch complex at Kennedy Space Center). The LCP Version 1 did not yet, however, include any specific named potential commercial suppliers in this category. Honeybee Robotics has been active in this area since 2015 or even earlier. Redwire is a more recent potential player, having developed technology for 3-D printing of regolith, which they intend to demonstrate using simulants at their Additive Manufacturing Facility (AMF) on the ISS. They have developed tools for lunar welding, microwave sintering of regolith, and manufacture of basalt fiber/regolith materials. SafeAI is also potentially part of the supply chain, having developed ways of automating heavy equipment for mining and Earth-moving.

Who Are the Potential Customers? The customers are likely to have a large overlap between the landing pads and roadway subsegments. They are the transportation providers of market sector 1, space agencies, and commercial private actors. The list of entities known at present includes NASA, ESA, Roscosmos, CNAS, JAXA, ISRO, CLPS Providers, ULA, Blue Origin, SpaceX, CSDC, and Arianespace. We have not been able to find the commercial arrangements, including pricing, that would apply for this kind of work. For the Mature Phase, a fully regulated Moon Port infrastructure steady state is envisaged, and the roadways will form part of a fully-regulated route structure, as part of a regulated Moon Traffic Management steady state. What do *you* think is the likelihood of this happening any time soon?

What Are the Likely Drivers and Constraints? The market drivers for the pads will be the number and traffic rates. And for the roadways, depending on length and expected traffic. It is unclear whether a toll-road system will be adopted. Regarding constraints, on the demand side, this will involve the developing lunar environmental regulations, including debris and dust mitigation measures. On the supply side, we await the development of the technology for sintering the lunar regolith to produce the berms, pads and roads.

How Does It All Fit Together in a Value Chain? It remains unclear at this time whether, and how, the building and maintenance of launch pads and roads will be subject to a centralized authority. It seems unlikely that a true operational launch pad, or roads, will be built during the Early Phase. The transition will therefore be from an *ad-hoc* process during the Early Phase, toward a regulated structure under a Moon traffic management model (presumed to be under the oversight of a Moon Port Authority). There may also subsequently be a final element of the value chain related to decommissioning and recycling of previous generation routes made obsolete and/or noncompliant. Can this really be a free-standing business?

How Can We Improve the Data? What research will be needed on an ongoing basis to improve the data for this part of the infrastructure? What will be needed in order to design and produce the follow-on systems to make it possible to move forward to the true commercial lunar marketplace? There are a significant number of experiments taking place both on Earth, and then subsequently, and relatively soon, on the lunar surface, about working with the regolith, and we can expect to gain knowledge from their outcome which will improve our projections.

A summary of this market sector is provided in Fig. 9.3:

Investing in these infrastructure-related market sectors requires that you carefully monitor what the respective involved governments are planning to provide, and their assumed funding and timescales. You will need to be able to separate the practical, funded, projects, and proposals from the wishful-thinking. And let's not forget that these infrastructure elements, as important and basic as they are, have not (yet) included using the hydrolysis of water ice to provide the basic oxygen needed for survival on the Moon without having to reply on supplies from Earth. That comes in the next chapter, when we really try to cut the cord with the home planet, and truly "live off the land" on the Moon.

Market Sector #5 Civil Engineering

Description
Building, Operating and Maintaining Launch/Landing
Pads and Roadways on the Moon.

Potential Customers
Government Agencies, All 10 Other Market Sectors,
Especially Transportation Providers.

Potential Suppliers Architecture/Engineering Firms,
Miscellaneous Hardware & Services Providers, Sintering &
Berm Construction Companies.

Drivers And Constraints
Regulatory and Environmental Norms, Traffic Rates,
Length of Roadway Network, Toll Pricing Feasibility, etc.

Value Chain Schematic

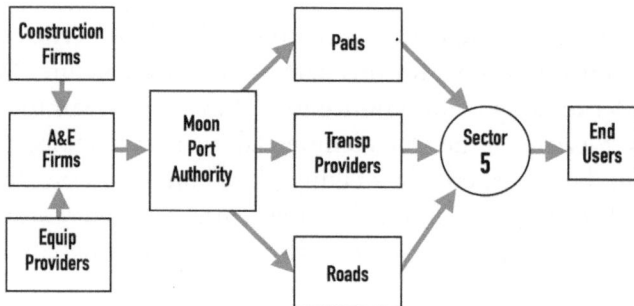

Fig. 9.3 Summary of LCP data for sector 5—Civil Engineering. (Credit: DW/MVA)

References

DARPA. (2023). *A framework for optimized, integrated lunar infrastructure.* Darpa. mil/news-events/2023-08-15

IOAG. (2000). Interagency Operations Advisory Group (of UN Office for Outer Space Affairs). https://www.ioag.org

Schmitt, H. (2021, August 18). *LSIC, New Science from Apollo.* In Lunar surface science workshop.

Scott, D. (2021, September). *LEAG, 12 action items for optimizing human exploration and lunar science in the next 5 years.* LEAG annual meeting.

10

Staying There

Just about everything we have discussed up to now as market sectors has depended pretty much upon governments as customers, at least with regards to the Early Phase of operations (that period up to 2030). So, our general public questioners, referred to in the Introduction, could quite rightly deduce that there is no truly commercial market on the Moon. They (we) could claim, with justification, that just about all the money discussed so far represents outgoings. Outgoings funded by themselves, the tax payers of the world. Sometimes, the government comes under cover of another name, such as academic research, but they know, and you know, that it all comes down to government money in the end. So, are they right?

Most reasonable people would I think be forced to agree with them, at least regarding the remainder of this decade. So, it is indeed misleading to suggest otherwise. Naturally, there are lots of potential commercial suppliers of products and services to and on the Moon, even during this period. And many of them can be expected to make a good sum of money, but they will be doing so in dealing with the government as their only true customer. But the task we have set ourselves, and the task undertaken by the volunteer analysts of the Lunar Commerce Portfolio, was to investigate the extent to which a permanent settlement on the Moon could generate a true commercial lunar economy, and so far in this account, that work has yet to be demonstrated, or at least investigated. It does seem to be quite a challenge. In this chapter, we address the remaining, and incidentally the least understood, lunar commerce market sectors of 6 thru 11. This is where we really begin to figure out what "living off the land" will mean when "the land" is the hostile lunar surface. Can we make it work? We know it is important that we try. We even need to

D. Webber, *Lunar Commerce*, https://doi.org/10.1007/978-3-031-53421-8_10

ensure we can deliver the basics, such as having oxygen and water available in sufficient quantities without recourse to deliveries from Earth.

We know that whatever new sectors are introduced into our analysis, they all presuppose that the above infrastructure (and indeed lunar tourism) sectors are in place. In describing these new elements, we shall in effect be relying upon the definition of the "Mature Phase." You may recall that the definition adopted, by consensus of the international analysts of the Lunar Commerce Portfolio, involved the phrase: "sustained by the Moon's resources, and not dependent for the necessities of life on a logistical supply chain of deliveries from Earth." What are the implications of this definition? What do we consider to be "the necessities of life"? The LCP group left that deliberately vague. But we have a pretty good idea. Like water. And food. And oh yes oxygen. Some of us need a coffee in the morning—but is it a true necessity of life? You see what I mean? Well, in reality, if we proceed to develop the Moon, we can expect that there will be a gradual improvement, from the Maslow basics (Maslow, 1943), toward the eventual presence of, say, a movie theater, games room, and bar for tourists, and casinos, and heaven knows what. There will certainly need to be habitations of various kinds, tourism hotels, restaurants, supplies and maintenance facilities, medical facilities, lunar agriculture, and fish farms. Tourists will in all likelihood want to experience lunar sports and will no doubt at some point express the need for appropriate facilities for, e.g., human flying (with 1/6 g it will be possible) (see, e.g., Leap of Faith, 1995). Governments will want research facilities for science and technology testing, and administration facilities. There will need to be traffic control and other regulatory facilities. As part of a long-term lunar economy, there could be manufacturing operations, which will take advantage of the hard vacuum and the 1/6th g that operates on the lunar surface. And mining. Mining will be an essential part of sustained living on the Moon. How else will we be able to provide for water and oxygen? There might also be the possibility of mining on the Moon to satisfy demand on Earth for certain rare materials, such as Platinum Group Metals (PGMs), rare earths, and maybe even He3 in the longer term, if the means of providing fusion power on Earth becomes practical. Metals could also be mined on the Moon for use on the Moon. And bearing in mind the concept of using the Moon as a gas station for Earth, we could manufacture rocket fuel from the lunar regolith and the ice expected to be discovered at the lunar poles in cold traps. There will be construction. There might be archival storage for Earth. There would be advertising and video businesses operating on the Moon. It is therefore going to be our task to assess the revenue generating possibilities of these operations independently, and when operating together, tie them to our assumed driving assumptions, crank the handle of the model, and find out what revenues would be generated under different circumstances. All of which, of course, will ultimately need to comply with whatever international and national guidelines are adopted for such matters as environmental protections (whatever that phrase might even mean on the Moon).

To make this possible, we refer back to our standardized approach, and lay out in turn what we can establish about each of the sectors 6 thru 11. By necessity, you will have noticed that as we move from market sector 1 to market sector 11, it gets harder and harder to come up with good data sources. That is to be expected. That is part of the reality check. We don't even know when these businesses will be operating. It may be just after 2030. But more likely not until nearer 2060. But in Version 1 of the Lunar Commerce Portfolio, we made a start, and will rely on future iterations to narrow the uncertainties. In some situations, we simply admitted defeat at this time, and left the matrix incomplete with an N/K indicator that work is needed in order to improve things for the next version. At least by using this model and these formats, we are able to identify where our data is problematic, and therefore where research is urgently required. It can form a wish-list for the MVA and Bocconi analysts, aided by the Lunar Commerce User Group, to guide their future work. Remember, also, that we were an international group, and searched for international source material, although at this stage of lunar development, despite contributions from all around the world, there may still seem to be a preponderance of US material. We may expect in future versions to find more contributions from China and India in particular. With that caveat, we continue to lay out the remainder of what we consider to be the potential lunar business market sectors.

Market Sector 6—Habitation and Storage

This might seem to be an obvious category. We'll have to live somewhere, right? Yes, but we do need to be careful in our definition to make sure nothing is omitted, and no double-counting occurs. So now we go to our regular format.

What Is the Formal Description of the Segment? We define "habitation" for our purposes in the LCP as a pressurized, protected environment on or under the surface, or in lunar orbit, where people can live and work. "Storage" is defined as being pressurized and unpressurized enclosed spaces for vehicles, equipment, tools, samples and stocks. During the Early Phase up to 2030, there are not assumed to be any lunar tourism facilities on the surface—only in lunar orbit (assumed to be at the Lunar Gateway facility). Only government habitation facilities will be needed on the surface, and during this Early Phase, the habitats will arrive with the inhabitants for each mission. During the subsequent Mature Phase, there will on the surface be more tourism-related habitats including restaurants, entertainment facilities, arts centers, bars, casino, and even sports facilities—built substantially from locally sourced materials. Among the government habitation facilities will be laboratories for the continuing scientific studies, for exploration and resource utilization experimentation. We remember that science and exploration were the original motivation and purpose of the space program, and it will continue to be important in this return to the Moon. It is possible that specialized storage will be needed for the commercial long-term business of archival storage. In lunar orbit, we assume some expanded habitat space for the lunar orbital space tourists, either attached to the governmental Lunar Gateway station, or as a separate commercial space station/orbital hotel facility. So, this sector is a real business. It is a real estate business. The actual building of the facilities is covered for our purposes under the rubric of Sector 8, Manufacturing (to avoid double-counting of demand). The services offered under this area of commercial business include lease and management of the surface and orbital facilities, the selling of real estate to private individuals, companies and government agencies, and hotel services to people for short and medium stays on the lunar surface and in lunar orbit.

Who Are the Potential Suppliers? There is in the LCP a substantial list of potential suppliers for this market sector. They include Kaiser aluminum, Valbruna Group, Arconic, Alfa Meccanica, Thales Alenia, Northrop Grumman, SpaceX, Yuri Gravity, and Space Tango. During the Early Phase, the list of companies who have expressed an interest includes Penguin Automated Systems, Orbit Fab, Planetary Shelter, Wasp, and Astroport Technologies, together with some national space agencies, such as NASA, CSA, JAXA, and ESA. Work within the LSIC organization has shown how habitations are able to be built using 3D manufacturing techniques, using simulated lunar regolith as raw material (e.g., LSIC meeting 25 June, 2021, Giulio Buscaroli of the Italian firm WASP demonstrated their modular printer for building called Crane WASP, which excavates the regolith *in situ* and then transforms and shapes to create the printed building).

Who Are the Potential Customers? They will include government agencies, universities, companies (such as Blue Origin) and private tourism companies. Among the commercial companies are Yuri Gravity and Space Tango, both planning to use the pressurized habitats for conducting scientific experiments. The customers will differ, depending on the kind of habitation or storage facility that is being offered. Some will buy the facility; others will lease. From the government perspective, all the international signers of the Artemis Accords are potential customers for some part of the lunar surface habitat volume. Among the commercial customers will be the space tourism operators who intend to provide a lunar space tourism hotel, and restaurants, bars, and ultimately casino, sports, and arts facilities on the lunar surface.

What Are the Likely Drivers and Constraints? The main driver for this sector will be associated with the ability to price in such a way as to provide an ongoing business case. Our current estimates are rather poor at this stage, and some high-quality market research is needed. The initial customers will likely be governments. Pricing will vary depending on whether the customers are long-term (greater than 6 months), or short-term. For the purposes of this preliminary study, prices were derived by adding a margin to an assumed cost (see LCP for details). The weekly cost per person was derived assuming the minimum cost will be the repayment of the capital invested, whether for orbital or surface facilities, assuming a ten-year time horizon (because of assumed private/public partnership). On the constraints side, the issue would revolve around the likely building rate, especially when the necessary outfitting was included beyond the basic 3D construction process. However, perhaps more significantly, there is the need to have a clear understanding within international law of some kind of ownership processes, particularly with respect to lunar surface facilities. It will be essential for those who will be owners, or who will use the facilities on the Moon for residential purposes, to enjoy, if not the full right of ownership, at least a protection similar to it, that allows prolonged use of the space. Work is still needed to evaluate how various design solutions mitigate against the likely lunar radiation environment. A NASA Langley study, reported at an LSIC meeting in June 2021, reported that shielding by regolith may be adequate, but that new technologies involving the use of a polymer overcoat of habitations could also be effective. Work goes on. The NASA Langley Principal Technologist, in an LSIC meeting August 20, 2021, pointed out the severe technology gaps involved in the outfitting of 3D manufactured habitats on the lunar surface. He noted that there would be a need to include power, lighting, HVAC, water, coolants, insulation, windows, hatches and interior furnishings. And that there are severe technology gaps in achieving this, particularly related to power and data cable line management (including micrometeor and radiation protection), piping/tubing line management (including joining, testing and repair), and penetration/sealing management.

How Does It All Fit Together in a Value Chain? We need to be careful to avoid double counting of demand potential. In the Early Phase, most habitations will be delivered direct from Earth to the Moon and assembled *in situ*. They will be designed and manufactured on Earth by manufacturers/systems integrators, and then transported to the Moon as cargo, possibly involving some astronauts in the final assembly and fitting out. There may also be the need for some radiation protection by a final layer of regolith which will require bulldozers on the surface (Market sector 2) to conduct the work. The government agencies will have oversight. In the true Mature Phase, we build habitats of various kinds (handled for our purposes under market sector 8) from bricks manufactured from the lunar regolith and/or via 3D printing of buildings. In this subsequent stage, there will be the need to separately have the buildings fitted out with such things as plumbing, pumps, cabling, CO_2 scrubbers, and furniture and lighting installed. The Sector 6 market opportunity is for the commercial real estate business of buying/selling and leasing of these habitats and storage facilities. The value chain for the Mature Phase therefore goes back as far as extraction of raw materials, and refinement and manufacture of materials for construction (both on Earth and on the Moon). So, this means lots of potential individual businesses. Both private developers and government will have oversight responsibilities. At present, it is not known what the lifespan of one of these habitats will be in the lunar environment (due to unknown effects of radiation, etc.), either in the Early or Mature Phase designs, and so it is difficult to estimate the timeframe for the maintenance cycles that will be required. Even the requirements for storage areas will need to take into account the need for humans to sometimes visit, so the requirements will not be trivial. It may be that it makes more sense to build habitats *below* the lunar surface (to protect from micrometeors, radiation, and even the widest extremes of temperature). In which case we should need to include an excavation process at the beginning.

How Can We Improve Our Planning Assumptions? What would it take? The problem here is that we are talking about so very far into the future. Market research can sometimes be helpful in these circumstances, but it will not be easy to create a statistically valid survey of appropriate people to answer demand-related questions about future needs for habitation and storage on the lunar surface.

A summary of this market sector is provided in Fig. 10.1:

And now we move to Sector 7.

Market Sector #6 Habitation And Storage

Description
Design, Building, Outfitting and Operating Facilities on The Moon for Government and Commercial Users. Also Bulk and Pressurized storage.

Potential Customers   Government Users Of Laboratories (NASA, ESA, Universities). Contractors. Tourism Companies (For Hotels, Restaurants, Sports Facilities). Other Market Sectors (3, 4, 5, 7, 9, 10).

Potential Suppliers   Kaiser Aluminum, Valbruna, Artonic, Alfa Meccanica, Thales Alenia, Northrop Grumman, Yuri Gravity, Space Tango, Penguin, Orbit Fab, Planetary Shelter, etc.

Drivers And Constraints   Pricing Model For Short and Long-Stay Inhabitants. Building Rate And Outfitting Technologies. Transition From Earth To Local Sourcing, etc.

Value Chain Schematic

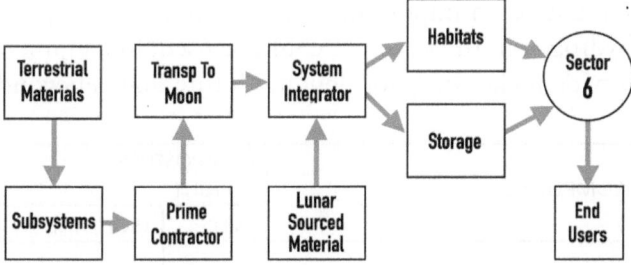

Fig. 10.1   Summary of LCP data for Sector 6—Habitation and Storage (Credit: DW/MVA)

Market Sector 7—Mining and Resource Extraction

We continue with our standard reporting structure.

Formal Description of the Segment. At last, we get to the key questions about the Moon with regard to mankind's possible future uses. What is the Moon made of, that would make it worth our while trying to harvest it? If it has been sitting there for 50 years since the last Moon landings, it cannot have any inherent value, surely, otherwise someone would have been back there already. Let's begin to answer these questions about whether it makes sense to mine. Starting with what is the formal description of the segment. It turns out that there are mineral resources on the Moon, but only those which are relatively rare on Earth might be worth mining, at least with regards to sending back the resource to Earth. So, it's probably *not* going to be about digging up the Moon to bring it back to Earth. The main rationale and incentives behind creating this commercial business sector are the potentially reduced costs for products intended for use *on the Moon*, or in lunar orbit, compared with having them sourced and transported from Earth. There are a series of source books that have considered the possibilities (e.g., Benaroya, 2010, Schmitt, 2006, Sivolella, 2019, Spudis, 2016, Wingo, 2004).

The following figures provide a summary of our current knowledge of what the Moon's made of. Figure 10.2 shows the Moon's mineral composition.

And Fig. 10.3 then provides the breakdown when reduced to its basic elements.

So, there may well be markets on the Moon itself, or in lunar orbit, for products found on the Moon, and these operations may be the most significant regarding lunar commerce. There is expected to be water in the form of frozen ice in those certain special category permanently shadowed regions near the poles. And the water could be used to provide oxygen, in addition to drinking water, and as an important part of manufacturing processes on the Moon. The hydrogen in the water ice could be used, in conjunction with the oxygen, to provide rocket fuel (for ongoing future flights beyond the Moon,

COMPOUND	COMPOSITION BY WEIGHT	
	MARE	HIGHLANDS
Silica (Silicon Dioxide)	45.40%	45.50%
Alumina (Aluminum Oxide)	14.90%	24.00%
Lime (Calcium Oxide)	11.80%	15.90%
Iron Oxide	14.10%	5.90%
Magnesia (Magnesium Oxide)	9.20%	7.50%
Titanium Dioxide	3.90%	0.60%
Sodium Oxide	0.60%	0.60%

Fig. 10.2   Average distribution of lunar surface minerals (Credit: Wikipedia – Geology of Moon)

ELEMENT	OVERALL COMPOSITION BY WEIGHT
Oxygen	45%
Silicon	21%
Aluminum	12%
Calcium	10%
Iron	6%
Magnesium	5%
Others	1%

Fig. 10.3 Average elemental composition of Lunar Regolith (Credit: The Chemical Engineer)

and for fueling return to Earth). In this market segment, we include resource characterization, extraction, processing and refining. During the Early Phase, this will largely be limited to experimental ISRU technology testing, including prospecting for ground-truth data, drilling, and sample analysis, and dependent largely on space agency funding. The aim would be to build a comprehensive map of the resource distribution and concentration in areas of interest. Limited resources would be extracted.

During the Mature Phase, by definition, there must be enough water and oxygen extracted and processed to sustain the life of lunar inhabitants, both on the surface and in lunar orbit, and to make the rocket fuel needed for the assumed mission cadence. Some of the regolith mining will have ensured that basic infrastructures on the Moon can be entirely shielded. Some extraction of Platinum Group Metals, KREEP/rare earths, and possibly He3, could be conducted for terrestrial use. Remote sensing data suggests that this resource is available, although ground truth (i.e., being there, on the spot, to check) has yet to confirm it, and it may not exist as ore bodies, making extraction a more difficult process. In our Scenario Delta analysis, where we assume sale of PGMs by export from the Moon to the Earth, our assumption was 17 tons of PGMs per year in that Mature Phase. The low gravity and high vacuum and extreme thermal environment imply the need for specially designed lunar-specific hardware and its maintenance. Regolith dust issues dictate special measures. Processing will include crushing, sorting, water/gas separation, electrolysis and metals separation. The findings of Clementine and LCROSS, and Chandrayaan-1 and LRO, have indicated that there could be 1 billion tons of water ice at each lunar pole. And the extraction and processing of this resource is the absolute key to sustainability of any kind of outpost on the Moon. We eagerly anticipate the findings from the next lunar spacecraft provided by a range of countries which will help refine our understanding of lunar resources beyond the information that was available to us in November 2022, when we conducted the LCP work.

Who'd have thought that water, plain and simple water, amid that "magnificent desolation," would turn out to be that precious resource that the Moon has been harboring, and which might be the basis of a new business opportunity? Ice mining on the Moon. Also, the lunar crust is found to be rich in iron, titanium, calcium and silicon, which can be a cheaper source of these minerals, for specific use on the lunar surface, than in bringing them up from Earth's gravity well to the Moon. The abundance of Helium 3 is greater on the Moon, having been embedded in the regolith over billions of years, and so might be a minable resource for terrestrial use in the distant future, if or when we learn how to create a workable fusion reactor on Earth. He3 in Earth's atmosphere is available at 7.2 parts per trillion (with a "t"), whereas on the Moon the concentrations might be as high as between 1.4 and 15 parts per billion (with a "b"). Iron is abundant to the tune of 14% per weight in the mare basalts, and Titanium content in some parts of the Moon is 10 times greater than on Earth. Silicon (21% by weight) would be available for manufacture of solar panels. Aluminum is available at over 10% by weight, and could form a source of solid rocket fuel.

In summary, we can say that the following lunar resources are the likely targets of lunar mining operations. Water, for life support, hygiene, cooking, and agriculture, is available at expected low to moderate volumes. The assumptions for the three Scenarios we used in the LCP were from 40 tons of lunar ice per year (scenario BETA) to 560 tons/year (Scenario Delta). Oxygen for life support and rocket propellant use is known to be available in very large volumes (hundreds of thousands of tons). Hydrogen is also used as rocket propellant, and can be obtained from the electrolysis of water, and might be available in moderate to large quantities (a few dozen tons). The regolith itself is ubiquitous and can be used for landing pads, roads, radiation shielding, etc. The metals can be used for manufacturing tools and equipment, pipelines, electricity grids, communications networks, etc., and are available in low to moderate quantities. PGMs are used for industrial purposes on Earth, and are considered increasingly as scarce resources (and one which has strategic sourcing implications, since much of it on Earth comes from China), as the demand has increased, particularly as a catalyst in the manufacture of electronic batteries and other components, with a potential lunar-sourced scale of around 400 tons. KREEP in general is used on Earth in the manufacture of electronics, clean energy, aerospace, fertilizers, glass production, and in making permanent magnets, although its quantities on the lunar surface are yet to be firmly established. The potential uses of He3 on Earth, beyond the fusion energy source, include cryogenic research, medical and quantum computing, and in particle colliders for subatomic research. We need ultimately to be able to decide, for each of the likely lunar resources, whether they are *just* resources (i.e., a concentration of naturally occurring solid, liquid or gaseous materials in or on the surface in such form that economic extraction of a commodity is regarded as feasible), or capable of being described as reserves (i.e., that portion of an identified resource from which a usable mineral or energy commodity can be economically and legally extracted)—thanks to Clive Neal of the University of Notre Dame for this characterization, provided at an August 2021 LSIC meeting.

Who Are the Potential Suppliers? During the Early Phase, NASA is providing, under its CLPS program, a number of exploratory robotic rovers with missions related to finding, drilling, and in other ways understanding the properties of the regolith for potential future mining purposes. There is a long list of potential suppliers in this segment, including OrbitFab, Galactic Mining, Safbai, Xiphos Technologies, Deltion Innovations, Honeybee Robotics, Masten Space Systems, Ball Aerospace, Asteroid Enterprises, Oxeon Energy, Teledyne, Sierra Space, Built Robotics, Austeer Engineering, WGM, Bechtel, Terraxis, Kilncore, Caterpillar, Blazetech, Lockheed Martin, Etiam, and Transastra. There are 21 individual companies and institutes listed in Annex B of the Excel model associated with the Lunar Commerce Portfolio, Version 1. A great deal of exploratory work has been conducted over the last several years, especially within academic institutions such as the Colorado School of Mines, into the peculiar problems associated with the process of mining on the Moon. In particular, they have been studying ways to remove and transport regolith from the dark polar craters in the search for water ice. To do this involves cooperation with other market segments such as Sector 4 which will need to provide adequate power in these remote locations to provide energy for the necessary heavy equipment to function. Lunar dust presents an enormous problem for equipment, especially robotic equipment, operating in the lunar environment—because of its inherent electrostatic characteristics and the un-weathered sharp nature of its constituent particles.

Let's take a look at a few examples from the potential supplier offerings to see what is likely to emerge. The Japanese iSpace rover has been contracted to search for water ice in the Marius Hill region near the lunar south pole as soon as 2024. Lunar Outpost is planning to soon land its Rocket M (Resource Ore Concentrator using Kinetic Energy Targeted Mining) on Masten's XL-1 lander. Honeybee Robotics has its TRIDENT drill on the NASA CLPS Viper mission, which is being delivered by Astrobotic's Griffin lander this year. Paragon is developing a system for water extraction known as Ionomer Membrane Water Processing (IWP). Transastra Corp is designing a Radiant Gas Dynamic (RGD) mining operation which would use RF, microwave and infrared radiation to heat permafrost and other types of ice deposits for extraction. The Israeli firm Helios has developed an electrochemical reactor capable of extracting oxygen, metals and silicon from lunar regolith by using molten oxide electrolysis. Austeer Engineering is working on a system for extracting water ice from the regolith using its GROWLER technology (Grading and Rotating for Water Located in Excavated Regolith). Aqua Factorum has a concept for producing 490 tons per year of metals (titanium, aluminum). Lunar Resources, Inc. would process 100 kg per day of regolith, thus providing 13 tons of oxygen per year and 15 tons/year of assorted metals (Iron, Silicon, Magnesium and Aluminum). Lots of action. Lots of possibilities. Fun for the engineers. But at present, most of it is funded by government in its various forms. Figure 10.4 provides some idea of what it will look like to be mining on the Moon. The image shows an early generation robotic mobile excavator.

Who Are the Potential Customers? The customers come under distinctive categories. First of all, there are the life-support needs for water and oxygen,

Fig. 10.4 Robotic lunar mining operation (Credit: NASA/Pat Rawlings)

implying everyone on the lunar surface, or indeed in lunar orbit. Then there are the customers needing product for their own use on the surface, such as the lunar manufacturing companies listed below in sector 8. And then there are the launch vehicle companies wishing to purchase propellant for their ongoing uses, including for returning to Earth (such as SpaceX, Blue Origin, Dynetics, and Intuitive Machines). Finally, there are the terrestrial customers for PGMs, Rare Earths, and possibly He3. Within the database of information collected during the creation of the Lunar Commerce Portfolio's Version 1, we identified 25 separate potential customers for propellant, some of whom needed it for multiple different purposes and vehicles. In addition, there were 7 identified potential life support customers, and 14 potential customers requiring the mined resource for *in-situ* manufacturing and construction work on the Moon. We did not, in this first version of the LCP, simply due to shortage and turnover of volunteer support, identify any potential customers for those mining products designed for terrestrial use.

In September 2020, a major LSIC event, the ISRU Supply and Demand Workshop, took place. Of course, due to then-prevailing covid restrictions, it had to take place virtually, via Zoom. But, nevertheless, it provided some excellent data, particularly on the demand for processed lunar regolith. Examples of presenters included Nick Cummings of SpaceX, who stated that his company would be needing "thousands of tons of propellant per mission."

A few examples in each category will serve to provide some idea of what the potential customers will be seeking. Blue Origin's New Glenn launch vehicle will be using LOX/LH2 for its second stage. Dynetics is proposing the ALPACA vehicle running on LOX/CH4 to be reused and refueled at the lunar gateway initially and then on the lunar surface. Intuitive Machines is planning its NOVA-C vehicle running on LOX/Methane and planning to use Moon-sourced propellant. Relativity Space will offer its Terran R vehicle, using LOX/CH4. SpaceX will need to obtain LOX/CH4 for its Starship Raptor engines. Among potential users of local life support supplies would be China with its International Lunar Research Station (ILRS) (China, 2023), which is planning for a sustainable presence on the lunar south pole, with stated intent to rely on *in-situ* resources. Among the manufacturing and construction firms who need local resources as part of their lunar surface operation, we can mention, e.g., Maana Electric (Lux), BIG Group (DK), and Lunar Resources (USA), all involved with manufacturing processes based on regolith.

What Are the Likely Drivers and Constraints? In analyzing this sector, a model published by the Canadian potash and uranium industry was used for guidance. It will be a particularly advantageous driver, if and when it is discovered whether ISRU processes are effective, and shoveling, digging, and transporting on the lunar surface do not contain hidden obstacles. Among the potential constraints would be the development of international and national regulations for lunar mining, involving requirements for dust and other environmental issues. Then, there would be the potential supply chain limitations involved in getting the required heavy equipment on to the Moon. At a November 2020 LSIC meeting, Dale Bouchet of the Canadian mining firm Deltion Innovations pointed out that at present we have little understanding of exactly how water deposition happens on the Moon, and that on Earth there are completely different processes which operate. For an industrialized process to be introduced on the Moon, such fundamental knowledge needs to be obtained before even a decision on technology selection for mining can take place. Subsequent research missions, such as those of the Chinese Chang'E 5 rover, will illuminate some of these matters. In August 2021, at an LSIC workshop event, the principal technologist from NASA Langley, Mark Hilburger, noted some of the remaining technology gaps which needed to be addressed before a fully sustainable lunar base could be established. He listed the need for dust mitigation for actuators, sensors, seals, joints. He included long-life lubricants and motors, and a dust-tolerant thermal control system. Then he added autonomous conduit and tubing installation, routing and connection technologies, and demonstrable laser- or microwave-sintering of regolith. And the capability to extrude molten or cementitious materials, and print-system cleaning and maintenance. Regarding the key resource of lunar water, we still have only limited knowledge, as was pointed out in a NASA Glenn report presented at an October 2021 LSIC meeting—we have no idea in what form it resides, we have no idea of the depth distribution, we cannot differentiate in radar data from water volumes and surface roughness. And in summary the information from current data sets is insufficient for reserve definition. Apart from that, we're fine, it seems.

How Does It All fit Together in a Value Chain? We need to apply the knowledge obtained from terrestrial mining operations. In the Early Phase, only limited experimental prospecting will be carried out, using the modified simple robotic rovers originally designed for the Google Lunar XPRIZE competition. But in the Mature Phase, we need to contemplate all the stages of similar work conducted on Earth, viz., exploration, sampling, testing, extraction, refinement, storage, and feedstock production, possibly going so far as beneficiation and sintering. Also integral with this work will be the need to use specialized rovers, both robotic and crewed. Almost all of this equipment, in both phases, will need to have been manufactured on Earth, and transported to the lunar surface. This is generally described as heavy industrial equipment, and so it imposes some clear requirements on the transportation providers, both in terms of cost/kg to the lunar surface, and indeed to the mass and volume capacity per delivery. Further interfaces are introduced by the need for waste treatment and data exploitation measures. Data has a value chain of its own at all stages of the lunar mining cycle, for mission planning, geological modeling, resource estimation and reserves quantification. This Mining sector's commercial viability is essential to our ability to have any kind of lunar economy. Can we make a rocket-fuel factory? Can we have an adequate supply of water (for drinking, agriculture and manufacturing purposes), and of course the all-important oxygen to breathe?

How Can Our Data Sources Be Improved? There are a number of investigatory bodies working on the "whats" and the "how-to's" of Moon mining—such as LSIC, LEAG, etc. Their ongoing work, and the new results from successive generations of lunar rovers (e.g., from CLPS, China, India), as it is progressively published in the public domain, will make it more certain the extent to which we can successfully and economically mine the Moon. We may perhaps stress the point that we are not simply trying to improve the science and to further characterize the resource content of the Moon in different locations, because there is some history of this topic, but rather we are moving to the engineering phase, and need to learn how to extract and process the lunar resource at an industrial scale in the harsh lunar environment. Of course, we should not forget that at present there are no clear international guidelines on how, or even if, this may be done.

A summary of this market sector is provided in Fig. 10.5:

Now, to Manufacturing.

Market Sector 8—Manufacturing

We use our standard database layout to summarize what is in the LCP.

What Is the Formal Description of the Segment? For this sector, we consider two main kinds of operation. On the one hand, there is the manufacture of materials for use on the Moon itself—mainly building materials, 3D-printed buildings, solar cells, etc. And then there are the specialized factories designed for making products for ultimate use on Earth. In neither category do we consider for our purposes manufacturing in lunar orbit. We are talking about the design and building and operating of the factories to manufacture the structural elements required for use on the Moon, or in special cases for terrestrial needs (these terrestrial-focused factories will rely upon the special

Market Sector #7   Mining

Description   Survey, Characterization, Excavation, Extraction, Processing & Refining of Resources for Lunar Use (Water, Oxygen, Hydrogen, Metals), and Terrestrial Use (Platinum Group Metals, Rare Earths, He3).

Potential Customers Water Ice for Life Support, Propellant and Industrial Use on Moon, for Gvt, Commercial Customers, Metals for Lunar Manufacturing. PGM's & Rare Earths for Terrestrial Customers. Other Market Sectors (1, 2, 4, 8, 9, 10).

Potential Suppliers Orbitfab, Galactic Mining, Safbai, Xiphos, Deltion, Honeybee Robotics, Masten, Austeer, WGM, Bechtel, Terraxis, Kincoe, Caterpillar, Blazetech, Lockmart, Etham, Tansactea, Ball, Asteroid Enterprises, Built Robotics, etc.

Drivers And Constraints Significant Potential Issues of International Regulation (Environmental, Dust, etc.). Emerging Knowledge of Ease of ISRU, Pricing of Resources, Especially those Intended for Earth, etc.

Value Chain Schematic

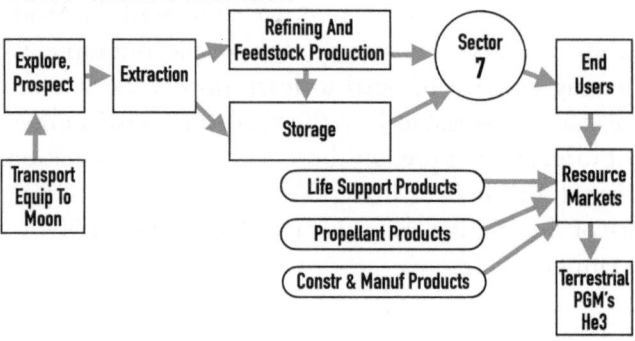

Fig. 10.5   Summary of LCP data for Sector 7—Lunar Mining (Credit: DW/MVA)

features unobtainable on Earth, such as the low-gravity environment and the high vacuum). In each case, we consider designing, building of facilities, and operating the factories to produce physical goods. During the Early Phase, there will be little of this activity. Almost all physical assets launched from Earth will be designed to be self-sustainable or require relatively little assembly, and be financed by governments. There will be engineering tests for bricks and concrete (lunacrete). There will be antenna and solar panel mounts. During the Mature Phase, there will be demand for all kinds of building materials made *in situ* on the Moon, and use of the 3D printing technologies to create ready-made buildings for the use of market sector 6. Bricks and concrete, glass, solar cells, batteries, microchips, electric motors, polymers, and ceramics will eventually be manufactured *in situ*. And they could be made into beams, sheet metal, pipes, wires, cables, trusses, gears, wheels, bolts, screws, and nuts. Construction and assembly will include habitation elements, hotels, research labs, storage units, storage tanks, space farms/greenhouses, optical telescopes, radio telescopes, radio antennas, power plants, and processing plants. For export to Earth, there would be those products that could not be manufactured on Earth in sufficient quality and quantity and which rely on low-g and hard vacuum, such as photoreceptors for copiers, some biotechnology products (pharmaceuticals, manufactured organs). Coming back to the reality check, we have to record that, so far, there has not been much engagement by these terrestrial manufacturers in this special category of lunar manufacturing. Probably, at this stage, there has not been enough outreach to make them aware of the possibilities. Getting the "Moon brick" businesses into shape might be a good place to start. Let's see what has in fact been happening.

Who Are the Potential Suppliers? The list in the LCP includes aerospace prime manufacturers, technology startups, research institutions and construction general contractors. Also, we have mentioned terrestrial companies wishing to use the lunar surface for part of their terrestrial manufacturing operations (such as photocopier manufacturers). In Annex B of the Excel model associated with the Lunar Commerce Portfolio, Version 1, we identify 60 potential contractors who might operate in this market segment.

Their interests and competencies cover a vast array of possibilities, and so it is hard for our purposes here to give examples which might be termed "representative." We shall see later, that in terms of revenue potential in this first version of the LCP, it seems that the most significant revenue source comes from the subsegment involved in building habitations in lunar orbit, although that may be simply a quirk of this first attempt to run the model. However, we do find among our potential suppliers at least two potential suppliers of those habitats—Sierra Nevada with their LIFE habitat prototype, and ILC Dover with their InFlex Lunar Habitat (ILH). We might single out some other examples of potential lunar manufacturers. Relativity Space is planning to provide 3D-printed structures on the lunar surface, and has received some NASA funding for this. And so does COSM (planning to provide an electron beam wire system for in-space metals printing), Thales Alenia Space (who have produced 3D-printed parts on the ISS), and Made-in-Space/Redwire (with its VULCAN advanced hybrid manufacturing system to produce high-strength, high precision polymer and metallic components). There are some sintering technologies (Pacific International, Astroport Space Technology,) some radiation shielding manufacturers (e.g., ICON and Liquifier), a great many design and architecture companies (e.g., Hassell, SAGA, BIG). Polimak is focusing on various conveyor systems optimized for the lunar environment (see presentation to LSIC June 2023), there are specialists in manufacturing underground, in lava tubes (e.g., 4th Planet, Stellar Amenities). And even manufacturers focusing on various food production needs on the Moon (e.g., Bake-in-Space, Cemvita, Zero-g Kitchen, and Space Zab). This is a very rich field of possibilities. Although very little has so far happened in a space environment. But it's early days, yet.

Who Are the Potential Customers? The customers on the Moon are many of the other market sectors (particularly 5, 6, and 10), i.e., those needing 3D-printed components, tools, spare parts, and solar cells. During the Mature Phase, customers may include space agencies of developing economies like Mexico, Brazil, Columbia, Chile, Egypt, India, Saudi Arabia, Australia, South Africa, and South Korea, among others. On Earth, depending on relative cost-structures, there could be some semiconductor users. Within the LCP Version 1, we identified some 51 potential customers of the various types of lunar manufacturing capabilities.

We can cite some examples. United States Structures is a lunar real estate development company planning in particular to provide structures for lunar orbit, including variable gravity environments. The Arch Mission Foundation is a nonprofit aiming to archive humanity's heritage for future generations, and which currently has a payload on Astrobotic Mission 1. Space Industries is a company intending to produce water and He3, and they currently have the world's largest 3D printer. SAGA is a Danish company of space architects. Relativity Space is an aerospace manufacturer. Orbit Fab will be a fuel delivery company in Moon orbit and then on the lunar surface. Maana Electric, from Luxembourg, is developing solar panel technologies. ICON is developing a full-scale additive manufacturing construction system. They all could potentially be customers for the suppliers of market sector 6—manufacturing.

What Are the Likely Drivers and Constraints? There are considerable unknowns at present about what can usefully be manufactured on the Moon for terrestrial customers. And so, it is a little early for any reliable demand calculations. This, in itself, is a constraint on market sector development. It suggests that there is need for some focused research into terrestrial firms who might benefit from low gravity, high vacuum factory facilities. A further potential constraint in regard to uses on the Moon itself might emerge from regulatory protocols that inhibit lunar surface manufacturing. In contrast, an effective driver of demand would be when/if it can be demonstrated that it is cheaper to make semiconductors on the lunar surface for lunar use, and even more so if the technology makes it possible to also be cheaper back on Earth, even when launch costs are included. In a December 2020 meeting of the LSIC Working Group on Excavation and Construction, it was proposed to mount a NASA Lunar Surface Construction demo in 2026, which would be a small-scale test (500 KG max) using a CLPS lander, which would test sintering and evaluate regolith simulant results. Some key architecture decisions are needed, which will have significant impacts on the way ahead. Most importantly, there needs to be a decision of whether habitats on the Moon are going to be *on,* or *under* the surface. There are potential suppliers with expertise in designing and building underground, often using robotic technologies (e.g., the Canadian firm Penguin/ASI) who have provided inputs to LSIC. Clearly, it makes quite a difference which approach is chosen. This underlines the fact that we are still a long way from lunar sustainability.

How Does It All Fit Together in a Value Chain? There is not expected to be much manufacturing during the Early Phase. But for the Mature Phase, the generic value chain involves the inevitable delivery of manufacturing equipment from Earth, although gradually some tools and spares will be produced by 3D printing of regolith materials. Then this sector relies upon the Mining sector to provide the appropriate raw materials needed for manufacturing. We do not yet know, but it seems likely, that most of the manufacturing on the Moon will be done for users on the Moon. But we assume that some manufacturers will emerge who wish to take advantage of high-vac and low gravity environments (as are available on the Moon) to augment their terrestrial manufacturing processes for their products destined for terrestrial customers.

How Do We Improve Our Data? Again, we address the important question about how we can progressively improve our planning assumptions for this sector. Perhaps the biggest unknown right now is the question about manufacturing for terrestrial uses. Are there indeed any companies on Earth who might find it to their advantage to consider a manufacturing component to take place on the Moon, because of its unique low-g and high-vac properties? This could be explored by researching various terrestrial manufacturing sectors via trade groups, etc.

A summary of this market sector is provided in Fig. 10.6:

Now for food production.

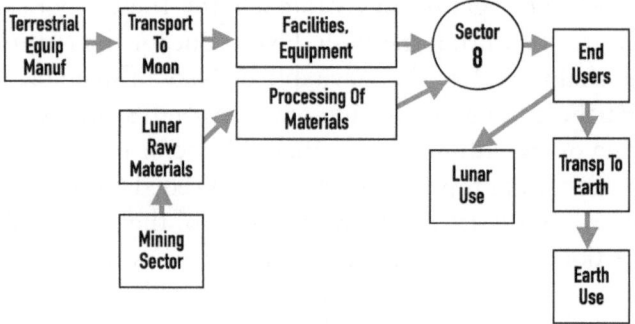

Market Sector #8　　　　Manufacturing

Description　Making Products on The Lunar Surface for Lunar Use (Building Materials, 3D Manuf Buildings, Solar Cells), and for Terrestrial Use (Products Requiring Low-g and/or Vacuum Environments).

Potential Customers　For Products On Moon – Other Market Sectors (5, 6, 10). 51 Identified in LCP. For Support of Terrestrial Manufacture – N/K – Maybe Xerox Corp.

Potential Suppliers 60 Technology Startups, Research Institutions, Contruction General Contractors, eg ICON, ILCDover, Relativity Space, Saga, BIG, Maana, etc.

Drivers And Constraints Level of Interest by Terrestrial Firms, Price Elasticities, Scale of Lunar Infrastructure, etc.

Value Chain Schematic

Fig. 10.6　Summary of LCP data for Sector 8—Manufacturing (Credit: DW/MVA)

Market Sector 9—Lunar Agriculture and Food

We know that it is possible. It was done in the Biosphere Experiment in Arizona. Now we adopt our standard layout of the assumptions we used in the LCP.

What Is the Formal Description of the Segment? Can you imagine what this would need to include? Remember, we are talking about being self-sufficient from Earth. Lunar agriculture is defined as productive activities undertaken using living things (plants, animals, cells, etc.) on the lunar surface. It is the science or practice of farming, including cultivation of the soil for the growing of crops and the rearing of animals to provide food, wool and other products. In the Early Phase, most of the activities will come under the heading of experimentation, usually taking advantage of deliveries of raw feedstock and growing facilities from Earth. There are of course amazing mass-efficiencies in bringing plant seeds and fish eggs to the Moon to be developed to full adults on arrival. All human trips to the Moon during this phase will however bring their own food supplies with them. Research will involve tailoring of crop varieties, using experiments such as those conducted under NASA's Veggie program. There are also experimental activities being carried out in India, China and Europe.

During the Mature Phase, the focus switches to self-sufficiency. Always assuming, of course, the explicit availability of adequate supplies of water at viable price levels. And oxygen, hydrogen, nitrogen, and energy. There will be a thriving lunar farm providing the raw product, and food processing and preparation areas. Experiments have already yielded some results so that we know the likely best species to be used. Vegetables will be expected to include arugula, roots, rice, pinto beans, tomatoes, and herbs. Fish would include tilapia, shrimp, and mussels. Among lunar-suited animals, it appears that rodents, goats and hens may be the most mass-efficient, as well as insect-derived foods. Yuck! Thankfully, researchers on Earth are already figuring out how to make this into delightful cuisine. The vegetables would be produced by both hydroponics and more traditional forms of agriculture. The fish farms would supply most of the protein. The animals would generally be husbanded to provide supplies of milk, eggs and meat. Assessment of needs, both on the surface and in lunar orbit facilities, will be based on RDI units (Required Daily Intake) and require 4.0 Kg/person/day (or only 1 kg of dry food) for 2000 cals/person/day. It is assumed that 11 kw per person is needed for food production, and 7 hectares of required area needed for 100 lunar inhabitants. The habitation segment will be the key driver of demand. There needs to be an interface with the bio-regenerative life support recycling system, where recycled human waste can be used as fertilizer (see Sector 10). Already, some astronauts in the International Space Station (ISS) have grown and eaten some experimental space crops, (as well as consumed their own processed urine as drinking water)—but we are still probably a long way from having the entire food source being provided locally (despite the graphic accounts of fictional Mars astronaut Mark Watney's efforts in Andy Weir's novel "The Martian"). There will be a need, in the preparation and processing phase, for automated farm machinery. We have some useful analogs to help in planning, including Antarctic settlements, nuclear submarines and space stations.

Figure 10.7 gives something of an idea of what part of the lunar food facility would look like.

Who Are the Potential Suppliers? Among those who have already been involved in discussions and experiments are Orbital Farm (of Toronto, Canada), Interstellar Labs Biopods (of Evry, France), Aerofarms (of Newark, USA), Infarm (of Amsterdam, Netherlands), CropOne (of Millis, USA), Square Roots (of Brooklyn, USA), and Bionetics (of Yorktown, USA).

Who Are the Potential Customers? Customers will be both on the surface and in lunar orbit and include governments (for their astronauts, and contractors, operating on the Moon), Research Institutes (for their lunar-based scientists), and Lunar surface tourism operators. For some purposes, it can be helpful to separate the various customers into those which are "permanent"—mainly government and Research Institute, and those which are short-term inhabitants (mainly lunar surface tourists, staying about two weeks at a time). Our assumptions for the quantification process involved between 10 and 200 permanent residents on the lunar surface in the Mature Phase, and between 5 and 130 in lunar orbit.

Fig. 10.7   Lunar greenhouse facility near the lunar poles (Credit: ESA/P.Carril)

What Are the Likely Drivers and Constraints? Among the constraints will be limits on availability of energy and water, particularly under lunar night conditions. There will also be the need to have an efficient and reliable biore-generative life support system (BLSS) throughout the lunar surface habitats (with 98% water recycling and 100% oxygen recycling). The drivers of demand are simply the numbers of permanent and short-term residents on the surface. Feedstock would initially be provided from Earth, during the Early Phase, at relatively little cost (particularly for seeds, insects, and fish eggs), and the future viability of the whole lunar operation, in the Mature Phase, depends, quite literally, on the ongoing farming processes. At this early stage, when data is so scarce, it is assumed that the price of food on the Moon is the same as that on Earth, plus the delivery costs, based on mass, of transport to the Moon. This becomes the target that must be beaten in reaching the Mature Phase. LSIC experimentation continues to establish the most likely candidate animals and plants. At the May 2021 LSIC meeting, one researcher presented data on Tilapia, noting that 2.9 kg/day would be needed for 12 people. These fish have an exceptional food conversion ratio of approximately 2:1, compared with beef cattle at 20:1. The researcher noted that a 21 m^3 grow-out tank would be needed. It was pointed out in the same contribution that several unknowns persist which must be addressed through testing, including the survivability of eggs or fry exposed to launch noise, shock and vibration environments, and the ability of fish to prosper at 1/6 g. So, much more work is needed.

How Does It All Fit Together in a Value Chain? In the Early Phase, most of the needs for food will be supplied direct from Earth and therefore need transportation, preparation and processing (ovens), distribution (including storage), consumption, and then waste recycling. For the Mature Phase, assuming that all the feedstocks were delivered during the previous stage, there will be no need for the transportation phase. However, we need to include an additional stage up-front, which is the step of food production on the surface. Provision will be made for handling and harvesting vegetables, fish, insects and animals. In the Mature phase, the customers will include lunar surface tourists, in addition to the government astronauts and contractors. Supplies will be needed to be produced and delivered to lunar hotels, bars and restaurants. Some research is being conducted into producing alcoholic drinks for lunar users. Regolith Reisling, anyone? This food industry, of course, is a well-known terrestrial business area—we are now contemplating replicating and establishing all the stages of the food business for the environment on the Moon.

How Do We Improve These Forecasts? A key subset of data is the unknown numbers of future lunar surface tourists (and the associated prices they must pay to experience living in a lunar hotel). Some high-quality statistically valid market research should be capable of refining the assumptions we have been using.

A summary of this market sector is provided in Fig. 10.8:

It is getting harder and harder to specify and analyze these remaining sectors. Now we move to an eclectic group of remaining submarkets, some of which might prove very hard to commercialize.

Market Sector # 9   Agriculture And Food

Description   Providing Food Supplies for Lunar Inhabitants from Vegetable Farming, Animal Husbandry, Fish & Insects, Including Processing and Preparation.

Potential Customers   Governmental and Commercial Occupants of Lunar Orbit Gateway and in Facilities on Lunar Surface. Other Market Sectors (6, 7, 8, 10).

Potential Suppliers Orbital Farm, Biopods, Aerofarms, Infarm, Cropone, Square Roots, Bionetics, etc.

Drivers And Constraints   Need to Meet Calorie Needs of Lunar Occupants. Need Initial Feedstock From Earth, Need to Meet Recycling Needs. Limits of Water, Energy,etc.

Value Chain Schematic

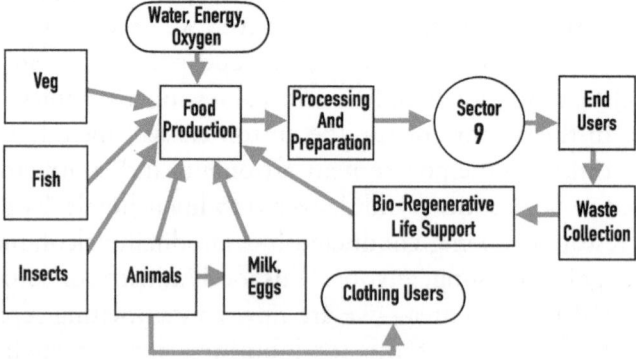

Fig. 10.8   Summary of LCP data for Sector 9—Agriculture and Food (Credit: DW/MVA)

Market Sector 10—Support Functions

What Is the Formal Description of the Segment? This sector represents a potential large need for lunar products and services. It includes the inhabitants of hospital/medical, governance, and lunar operations centers. And, furthermore, in this category we have included the management and operation of all recycling and waste management activities on the Moon. It is assumed that the hospital would require hyperbaric and centrifuge capabilities. In the Early Phase, it is assumed that no independent facilities for waste processing and recycling, beyond those in the arrival spacecraft, will exist. In the Mature Phase, the waste will be produced at scale, requiring several categories such as grey water, human and agricultural waste, and industrial waste processing. There may also be the need for nuclear waste disposal. Emergency fire/rescue/security services will have important functions during the Mature Phase. Eventually, there will be a need for cemetery services, schools, and many of the other supporting elements that we take for granted here on Earth. It is assumed that the funding for these Support Functions will be from government during the Early Phase, and may possibly contain some commercial elements during the Mature Phase. For instance, will education be provided as a commercial proposition? There will, we may be sure, be an essential general provision lunar supplies store/depot, where lunar inhabitants can come to obtain food, supplies, and have equipment maintained, serviced and repaired. We do not, in Version 1 of the Lunar Commerce Portfolio, include lunar orbit users for this market sector. But that supply store sure looks like a good proposition—running a lunar "Home Depot" store business.

Who Are the Potential Suppliers? In the Early Phase, all of this support activity would be funded by government. However, in the Mature Phase, there would be opportunities for entrepreneurial entities, wishing, for example, to establish and operate the lunar supplies store, including its equipment servicing department. In this, our first attempt at systematic data collection, we were not able to yet identify any specific potential suppliers for this market sector.

Who Are the Potential Customers? The customers would largely be governmental, especially in the Early Phase, although there would be a wider need for the medical and emergency service categories. And users of the supplies store would include anybody operating on the lunar surface.

What Are the Likely Drivers and Constraints? The effectiveness of the lunar surface supplies store will depend on the supply/demand balance that can be established. It is too early to be able to predict price levels for the various supplies, except that we know it will depend on the costs of delivery from Earth, which will need to be significantly reduced in order for heavy equipment to be delivered. Recycling and waste management will probably have to operate following some centralized agreed international protocols and interfaces which must be agreed. The good news on the "drivers" front is that an initial set of heavy equipment, once delivered, can support both the original small number of lunar inhabitants and the presumed increased numbers under the Mature Phase assumptions.

How Does It All Fit Together in a Value Chain? Although it is difficult right now to estimate the numbers of eventual users, and the frequency of their needs for these services, we can be sure that certain equipment needs to be delivered, ideally during the Early Phase, so that the emergency, medical, and waste disposal services can begin to operate, as the numbers of lunar inhabitants increase during the Mature Phase. Also, we do not know now, but we can assume that there will emerge, various regulatory requirements related to environmental protections on the Moon, and they might need, for example, a means to travel about on the surface to monitor activities. Also, included in this sector will be the all-important lunar supplies store, which will require its own supply and value chain. We may need to store nuclear and hazardous materials, a process which will bring its own associated components to be managed.

How Can We Improve Our Assumptions? Because we are looking so far into the future, it will be hard to refine these numbers very soon. Running a trading post on the Moon, including providing spares for equipment servicing, and other needs of lunar pioneers, will not be easy to achieve, or to plan for. Nevertheless, this will ultimately be a good measuring point for assessing the viability of lunar commerce itself. If this frontier store can make it, then there are good prospects for all the lunar commerce businesses.

A summary of this market sector is provided in Fig. 10.9:

We have arrived at our last market sector, as laid out in Fig. 6.1.

Market Sector # 10   Support Functions

Description   Operation of Moon Base Supply Store. Operation of Waste Management Process. Provision of Emergency Services (Rescue, Fire, Medical Support).

Potential Customers All 10 Other Market Sectors. Governments for Registries, Traffic Mgt, Zoning, Nuclear Waste Management

Potential Suppliers   N/K

Drivers And Constraints Numbers of Lunar Inhabitants, Regulatory Requirements, Need for Medical Facilities (Hyperbaric, Centrifuge), etc.

Value Chain Schematic

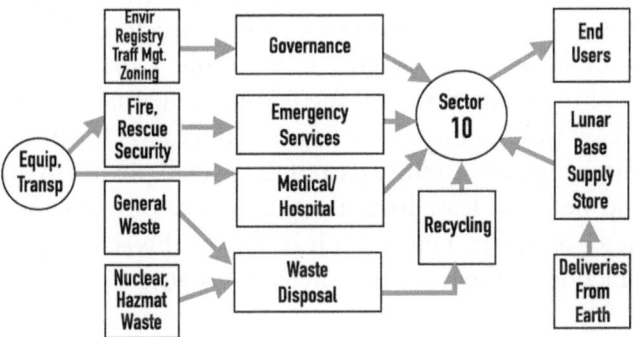

Fig. 10.9   Summary of LCP data for Sector 10—Support Functions (Credit: DW/MVA)

Market Sector 11—Other Commercial Opportunities

What Is the Formal Description of the Segment? This category was introduced to encompass "everything else," and so will be open to include future market sectors which we cannot imagine right now. It makes it possible to ensure that the market sectors of the Lunar Commerce Portfolio are indeed all-inclusive, by definition. At this early stage, it is unclear what new commercial opportunities might be involved, but a few are suggested, such as advertising, movie-making, and possibly the creation and management of data and financial products. Other ideas include possible storage of terrestrial archives of various sorts. But we are getting into the wild blue yonder as far as reality checks go. This is at present really just a "place holder" for future possibilities within the Lunar Commerce Portfolio model.

Who Are the Potential Suppliers? At this early and undefined stage, it is not possible to provide specific lists. Let's just call it TBD.

Who Are the Potential Customers? There may be customers both on the Moon and on Earth for this final undefined category, but it is not possible at this stage to provide specific names, or even categories, of users. So, again, let's just call it TBD for now.

What Are the Likely Drivers and Constraints? You know the deal. This category is so unspecified that it is not possible at this early phase to identify the drivers and constraints.

How Does It All Fit Together in a Value Chain? The chart in the summary sheet for this sector provides a generic concept, which is all that can be done at this stage of lunar development. It is simply a place-holder for future upgrading, and to ensure that nothing is forgotten. And, clearly, it does not indicate any easy ways to improve the uncertainty level for this sector. It may have to remain a high uncertainty part of the overall picture for a few decades, with the good news (at least from the perspective of our task of estimating lunar commerce revenues) being that it is only, at this stage, going to contribute a small fraction of the lunar commercial revenues.

Improving the Data? Notice that we have deliberately kept a space for the "etc" category. This represents all the unknown future lunar commercial activities, both on the surface and in lunar orbit, that we have not thus far been able to identify. But which remain as hopeful possibilities. When Sputnik 1 first went into space, who back then could have imagined all of the businesses that have emerged on Earth as a consequence of that first venture? The same must be true of what will take place on the Moon—and so we have kept a placeholder for it, even though we can't know how its value chain will develop beyond this initial indicated box on the chart. Almost by definition, this market sector will only emerge during the Mature Phase.

Although at this stage of only Version 1 of the Lunar Commerce Portfolio, we know so little, we nevertheless provide a summary of this market sector, to the now familiar format, in Fig. 10.10:

Market Sector # 11   Other Commercial

Description   Unknown Future Commercial Opportunities,
Data and Financial Products, Advertising, Movies, Lunar
Archival Storage, Funeral, etc.

Potential Customers   TBD

Potential Suppliers   TBD

Drivers And Constraints
Emerging Role of Moonport Authority, Emergence of
Lunar Financial Products, Regulation, etc.

Value Chain Schematic

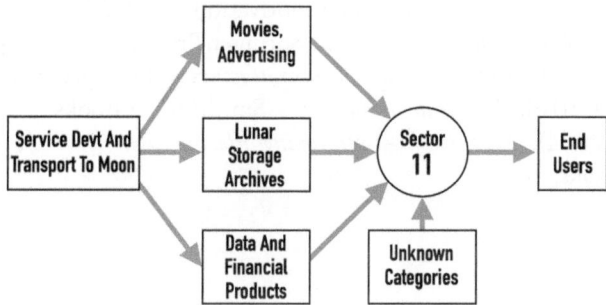

Fig. 10.10   Summary of LCP data for Sector 11—Other Commercial (Credit: DW/MVA)

So, do you now see what kind of activities might happen on the Moon? It is not so far-fetched at all, and is a logical extension to our lives here on Earth. Do you now appreciate how different it would be from the Apollo era? This would be beyond the imagination of Yuri Gagarin and the first astronauts. And you might even have an investment interest. And we know for sure that the further exploration of the Moon and the rest of the solar system will not, indeed cannot, happen without the development of the lunar commerce businesses. And you can play a part—how cool is that?

In this chapter, we have explored those opportunities for investment which at present are the least well-defined, but which might initially offer big rewards for the associated high risk. Certainly, you should have realized by now that investments in these "Sustainability" market sector areas should not be done in hopes of an early quick return. However, you do now have at your disposal a pretty good initial look at the sectors' respective value chains, so you might want to consider where you could jump on board when ready, with a view to possible long-term returns. Also, if your focus is perhaps more aligned with regulatory matters, whether at a national or international level, you have now seen the full range of possibilities that are being considered to take place on the lunar surface.

References

Benaroya, H. (2010). *Turning dust to gold.* Springer/Praxis.
China. (2023). *ILRS.* En.wikipedia.org/wiki/international_lunar_research_station
Leap of Faith. (1995). NASA artwork by Pat Rawlings/SAIC, nasa.gov/pdf/591752main_science-sports.pdf
Maslow, A. (1943). A theory of human motivation. *Psychol Rev.*
Schmitt, H. (2006). *Return to the moon.* Springer/Praxis.
Sivolella, D. (2019). *Space mining and manufacturing.* Springer/Praxis.
Spudis, P. D. (2016). *The value of the moon.* Smithsonian Books.
Wingo, D. (2004). *MoonRush – improving life on earth with the moon's resources.* Apogee Books.

11

Revenue Potential

The Lunar Commerce Portfolio Version 1, which was made public in November 2022, contained both a detailed documentation of assumptions, as you have now seen, and also an Excel model and its associated data, which was used to compute the revenues attributable to each of the market segments. Let's now use that same data in this primer to present to you the revenue potential from lunar businesses as we have defined them in the previous four chapters. This, in effect, is the "big reveal" of this book. As you would by now have realized, this information will be presented in the form of averaged annual revenue totals. Both for the Early Phase (from now until 2030), and for the eventual Mature period, whenever that is finally achieved. There will be no information for the intervening years (of unknown duration) during which there would be some kind of combination of the businesses described for the Initial Phase, and those for the Mature Phase. There will be both lunar surface and lunar orbit sources of demand. We are, of course, engaged in crystal ball gazing, and therefore do not know which of the markets we have proposed will become a commercial reality over time. But our work will hopefully provide some guidance and direction to the eventual creation of a lunar economy, and will offer some early order-of-magnitude estimates of potential lunar commerce revenues.

We shall also be able to show you (in the next chapter) how these revenue forecasts are heavily dependent on the external factors described by the Scenarios explored. So, for the moment, these will be the best center-lined available revenue forecasts for lunar commerce. Best in the sense that they have been derived from a set of publicly known assumptions. Best in that they are aggregated from collecting together all possible future market sectors. Best in that no double-counting has been allowed. Best in that they take into

D. Webber, *Lunar Commerce*, https://doi.org/10.1007/978-3-031-53421-8_11

account both the more recent findings of Moon resources, and the much lower potential costs of access. Best in that demand estimates and no hype have been included. Best in that this is all incorporated in a model which will allow for further iterations as new data emerges.

But not best in any sense of "perfection" or "precision." We shall note that there are some missing elements, where it was not possible for the volunteer analyst team to come up with public domain information in the period leading to November 2022. In having directed this work, I of course accept responsibility for its shortcomings (and offer an investor warning not to rely on the findings to make investment decisions without considering other possible sources). I do believe, however, that it represents a valuable basis for understanding and characterizing a future lunar economy. There are also some areas where it was clear that the data used was pretty weak. That, in itself, was an important finding. There is a lot of missing or weak data, and therefore, anyone else who puts forth anticipated demand figures for lunar commerce must also have faced the same problem, even if they did not necessarily state that fact. The uncertainties in the findings are a consequence of the real unknowns at this stage of the lunar development process. We have used the best available public domain data in reaching these findings. This work of the Lunar Commerce Portfolio, and its associated model, is designed for ongoing improvement. The Lunar Commerce and Economics Working Group of the Moon Village Association is committed to this continuing improvement. And, in fact, they have subsequently, in February 2023, made this case in presenting the current work to the UN's Committee on the Peaceful uses of Outer Space in Vienna (Gautel, 2023). The Moon Village Association, furthermore, signed a Memorandum of Understanding with the Space Economy Evolution Lab of the Business School of Bocconi University in Milan, Italy to work together to achieve these improvements, and in particular to understand how international regulatory decisions will affect the revenue outcomes (Bocconi, 2022). Also, following the presentation and free public release of the Lunar Commerce Portfolio Version 1 in Los Angeles in November 2022, an independent Lunar Commerce User Group (LCUG, 2022) was set up to engage with the model and its assumptions in order to continue to improve its findings. All of this activity will result in new updated versions of the LCP in the years ahead. Hopefully with much improved assumptions and consequently reduced uncertainties and risk. But I cannot guarantee that these updates will also be offered for free!

We have mentioned how we took into account the way in which the revenue outcomes would be impacted by external factors (Appendix B). And how we organized those factors into four representative scenarios to help us understand the kind of resulting impacts. We also provided detailed examples for two of them in Fig. 6.3. For convenience, below is a summary of the main elements of all four scenarios evaluated in Version 1 of the LCP:

Scenario Alpha—"Sorties": A continuous campaign of single missions to various locations around the Moon is assumed. Some lunar tourism will take place, but in lunar orbit only.

Scenario Beta—"Research Stations": One or more facilities similar to McMurdo Station in Antarctica are assumed. Lunar tourism is assumed both in orbit and on the lunar surface.

Scenario Gamma—"Sustainable Community": At least one permanent human presence on the Moon self-sustainable with the necessities of life is assumed (by our definition). Lunar tourism will be taking place both in orbit and on the surface.

Scenario Delta—"Resources for Earth": The Moon is fully open for business. Several government and private autonomous bases are assumed. Exports from the Moon of PGMs, He3, etc. will be taking place. Lunar tourism operates both in orbit and on the surface.

We now present the findings. How about a change of approach? Let's cheat a little, and start with the answer, which is normally in the back of the book. Rather than working through all the individual sector details, we are going to check out the Big Picture, the overall revenue outcome from all markets. This will be our first chance to get an inkling of where the big bucks will be coming from, and, equally important, which sectors do not seem to be very significant in the grand scheme of things. After we have done that, then we can go back and investigate which assumptions in each sector were the key to the findings. Consider Fig. 11.1, which shows the summarized results.

First of all, we look at the Early Phase. As you recall, this represents the sum total of governmental expenditures toward the return to the Moon, which might be compared with the annual budgets of the national space agencies reported in Chap. 4. You will see that the general level of business, expressed as an average year value, is at most $4 B. It hardly varies between the various scenarios. Continuing the Early Phase analysis, we found that the split between governmental and commercial revenues is approximately 50/50%, and furthermore, only market sector 1 (Transportation to/from the Moon) is generating revenues.

We now look to the Mature Phase, where all the action is. The totals vary between the various scenarios in ways we shall explain in Chap. 12. However, the big picture is that the annualized total revenues during the Mature Phase

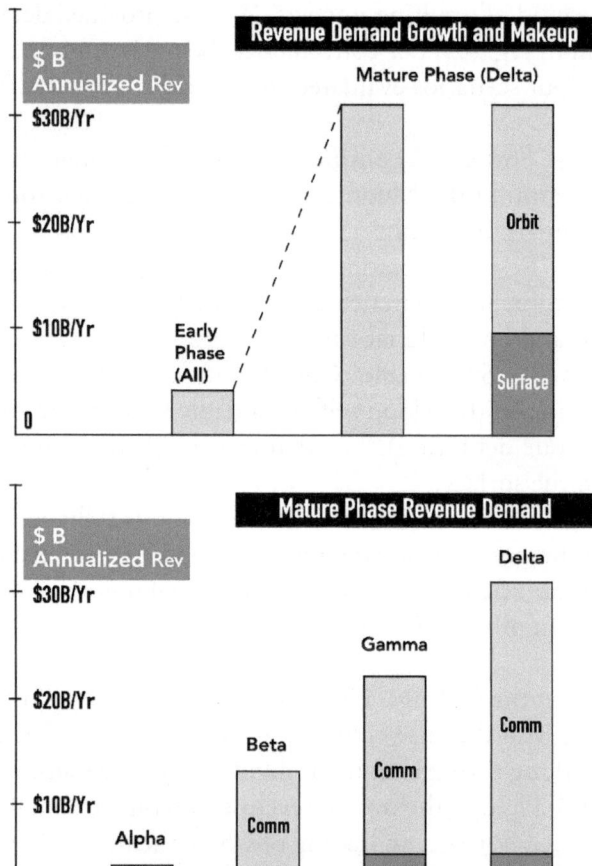

Fig. 11.1 Lunar commerce demand projections for Version 1 of the lunar commerce portfolio. (Credit: DW/MVA)

can reach about $31 B, with over 80% coming from commerce (Scenario Delta). Thus, all the growth between the two time periods is generally due to the new commercial market demand projections (the absolute governmental contribution is not assumed to increase significantly—it grows from $2 B in the Early Phase to at most $5 B in the Mature Phase). The main market sectors driving the demand are sectors 1, 6, 7, and 8, viz., Transportation to/from Moon, Habitation, Mining, Manufacturing, with all the other markets proving to be relatively insignificant with our current state of knowledge. And don't forget that lunar tourism revenues are explicitly included and spread across those market segments. We also determine that the split between lunar orbit and lunar surface revenues, using our assumptions, is about 40% from surface markets, and 60% from lunar orbit markets (Scenario Delta data), which admittedly seems rather surprising. We'll need to figure out why.

So, let's dig deeper. Although the full detail is available in the Lunar Commerce Portfolio, Version 1, we shall for this primer merely look at those four sectors that contribute 98% of the total revenues, to investigate which assumptions were key to the generated revenue results. We need to make this information transparent, so that subsequent updates will be possible. Also, we need to know if we made any really bad assumptions! So, here is the calculation process for each of the big contributors in turn. Please review for any flaky assumptions. Figure 11.2 provides the summary result for this first version of the LCP.

Although the full LCP contains all of the detail for *all* of the scenarios, we shall for ease of comparison just here use the full-up, or Scenario Delta, values. The Lunar Commerce Portfolio Version 1, when issued, contained a massive Excel model as described in the Fig. 6.4, with a first attempt at loading it with data, to the extent that data was available in the public domain at the time (November 2022). We should also note, for the record, that just about every data item had a range of values associated with it, and so decisions had to be made about whether to use the high end of the range, or the low end, or somewhere in the middle. So, we should not expect our ultimate revenue findings to be a precise number, but rather a cloud of possibilities, and the "true" value for lunar revenue demand would exist somewhere in that cloud. In future iterations of the model, it will be possible to be more systematic about which ends of a data item spectrum should be associated with which ends for another item. In general, of course, the higher the price the lower the demand, but future analysts will need to determine price elasticity of demand data for each sector in order to do this correctly.

With all those caveats, we now set out to explain the main calculation assumptions behind the findings that the $31 B/year revenue potential in the Mature Phase, for Scenario Delta, consists of $5.4 B/year from Transport to/from Moon, $12.5 B/year from Habitation and Storage, $4.3 B/year from Mining and Resource Extraction, and $8.6 B/year from Manufacturing (it will of course be a task for future analysts to review all the assumptions for the other 7 market sectors, to understand why the results for these sectors were so insignificant. Is this inherent to the lunar domain, or were there any miscalculations in the Excel data entries?). We now address each of the significant sectors in turn. After all, this is what we have been wanting to determine— which markets will prove to be revenue earners on the Moon, and on what assumptions has this finding been based.

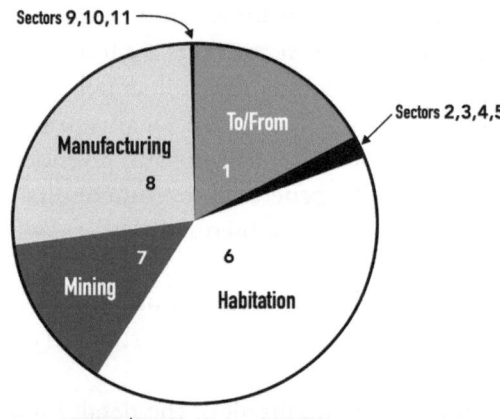

Fig 11.2 Lunar commerce revenue contributions by market sector, in the Mature Phase, for Version 1 of the Lunar Commerce Portfolio. (Credit: DW/MVA)

Sector 1 – Transp To/From Moon	$5.4 B/Yr	17%
Sector 2 – Transp On The Moon	$ 0.2 B/Yr	1%
Sector 3 – Comms and Nav	$ 0.01 B/Yr	0%
Sector 4 – Energy and Power	$ 0.3 B/Yr	1%
Sector 5 – Civil Engineering	$ 0.05 B/Yr	0.1%
Sector 6 – Habitation and Storage	$12.5 B/Yr	40%
Sector 7 – Mining	$ 4.3 B/Yr	14%
Sector 8 – Manufacturing	$ 8.6 B/Yr	27%
Sector 9 – Agric and Food	$ 0.04 B/Yr	0%
Sector 10 – Support Functions	$ 0.05 B/Yr	0.1%
Sector 11 – Other Commercial	$ 0	0%

Was this general result a surprise at all? It for sure took us a lot of work to get here. It's hard to say, because it's never been done before. But it does not seem unreasonable in any obvious way. It is what it is: the outcome from all of our data, assumptions, and understandings. This is the result from our first use of the Lunar Commerce Portfolio model. We can certainly live with it for now, as the overall outcome of our work. It will get harder to maintain our confidence level as we dig deeper to seek greater understanding within each sector, and it may reveal areas where more work is needed. We have four main sectors to explore, seeking this greater understanding (1, 6, 7, and 8). But don't forget that, although it does not have a dedicated segment of its own, under our stated assumptions it is lunar tourism which underlies much of what we are seeing. It creates much of the need for the four highlighted sectors discussed below.

Sector 1 Transport to/from Moon

This sector contributed $5.4 B/year in the Mature Phase, representing 17% of the total. The total of $5.4 B/year is the result of the addition of five subsectors as follows:

People	$1.6 B/year. This, in turn, comes from following the lunar space tourism assumptions included in the Appendix, and then assuming Starship deliveries to the lunar surface.
Cargo (nonpropellant)	$1.0 B/year. Again, Starships are assumed delivering 100 tons at $10 M per launch.
Propellant	$0.9 B/year. It is assumed that a Starship delivers 560 tons of lunar ice/year, and 17 tons of PGMs.
CLPS-type cargo missions	$0.4 B/year. Here, the assumption is an ongoing 2 missions/year with the government taking 75% of the capacity. Astrobotic published prices have been used.
Other small cargo landers	$1.5 B/year. In this category, we have assumed one iSpace and one Airbus mission/year for the use of "other" governments, and one iSpace mission per year for commercial use.

So, Starship is the key. Its success is critical to this whole endeavor, particularly if we can believe the lunar space tourism prices and forecasts that we have been using.

Now, we look at the sector which generated the biggest share of the Moon business revenues. Let's dive in to find out why.

Sector 6 Habitation and Storage

This sector contributed $12.5 B/year in the Mature Phase, representing 40% of the total. The total of $12.5 B/year is the result of the addition of four subsectors as follows (with a further five subsectors identified but not quantified in Version 1):

Habitation—Long-term stays (> 6 months) on surface (Government)	$0.4 B/year. Price assumed $660 K/week/astronaut. Based on assumed capital cost of lunar base, divided by a ten years' time horizon, plus 23% margin.
Habitation—Long-term stays (> 6 months) in lunar orbit (Government)	$0.9 B/year. Price assumed $1.3 M per week per astronaut. Based on assumed capital cost of lunar orbit space station Gateway, divided by ten years' time horizon, plus 23% margin.
Habitation—Short-term stays (< 6 months) on surface (Commercial)	$0.2 B/year. 1 tourist paying $290 M per ticket for two weeks in lunar surface tourist hotel. The $290 M price developed from the NASA contract award of $2.9 B and dividing by assumed 10 astronaut tourists in a Starship.

Habitation—Short-term stays (< 6 months) in lunar orbit (Commercial)	$10.9 B/year. 130 assumed lunar orbit tourists paying a ticket price of $80 M each for a 2-week stay.
Storage, etc.—none included for this Version 1 LCP issue.	N/K

These results seem, in hindsight, to be a bit out of balance, but it simply flows inevitably from the assumptions on lunar space tourists and their ticket prices. These lunar tourism calculations were made, we note in retrospect, before the loss of the Titan private tourist submersible, which event might have some repercussions to space tourism ventures. We don't know yet. Now we move on to look at mining.

Sector 7 Mining and Resource Extraction

This sector contributed $4.3 B/year in the Mature Phase, representing 14% of the total. The total of $4.3 B/year is the result of the addition of three subsectors as follows (with a further two subsectors identified but not quantified in Version 1):

Water Ice Mining for Propellant	$4.2 B/year. Assumes 1.1 M Kg needed (i.e., 10 Starships at 100,000 Kg per Starship), priced at $3600/Kg.
Water Ice Mining for Drinking	$0.02 B/year. Assumes 5300 Kg needed (with 80% recycled) and each astronaut or tourist using 0.12 Kg/ day.
Water Ice Mining for Lunar Agriculture	$0.04 B/year. Assumes 12,000 Kg of water needed for vegetables, aquaculture and animals.
Metals Mining for lunar use	N/K
PGM's Mining for terrestrial use	N/K

These results seem on the face of it to be reasonable, but of course we need more work to fill in the missing subsectors. Now for manufacturing.

Sector 8 Manufacturing

This sector contributed $8.6 B/year in the Mature Phase, representing 27% of the total. The total of $8.6 B/year is the result of the addition of four subsectors as follows (with a further three subsectors identified but not quantified in Version 1):

Building Habitations in Lunar Orbit.	$6.4 B/year. It assumes 3 orbital stations in lunar orbit at $32 B each, and amortized over a 15-year lifespan.
Building Habitations on Lunar Surface	$0.5 B/year. This assumes 4 lunar bases constructed at $1.7 B/base, and divided by the assumed 15-year lifespan.

Building Agricultural Production Facilities on surface	$0.1 B/year. This is calculated based on an assumption of 25% of the cost of the associated base, priced as above.
Building Agricultural Production Facilities in lunar orbit	$1.6 B/year. This is also calculated on an assumption of 25% of the cost of the associated lunar orbital station, priced as above.
Making Building Materials for use on Moon (bricks, solar cells, etc)	N/K
Manufacturing for Terrestrial Use—Biotech	N/K
Manufacturing for Terrestrial Use—Other low-g and high vacuum)	N/K

This sector's results do seem to be a bit problematical now that we are digging deeper, especially with regard to the lunar orbit habitation assumptions. We need more work to review that, and also to include the missing subsectors.

We have now seen all the intimate assumptions behind this $31 B/year revenue LCP forecast. Maybe it was not perfect, but I believe it is a good basis for moving forward. We know the uncertainties are high, and will discuss that in the next chapter, but at least you now know what was assumed by this Working Group of international volunteer analysts, working over a 2-year period (during an international covid epidemic!) to produce the Version 1 Lunar Commerce Portfolio. Because of the inherent uncertainties, it makes no sense to offer too precise a set of findings, and so we have rounded the values. It is even a bit unrealistic to insist on $31 B/year revenues, rather than, say, $30 B/year—but it provides an integrated whole of the (rounded) sector revenue values, which all add up to $31 B/year. We have developed what can be considered a Rough Order of Magnitude (ROM) estimate. So, we stick with that value, for practical reasons, while realizing that we do not really have much confidence in numbers expressed more precisely than about $1 B/year. Out of complexity come some simple realizations. We identified and analyzed as many as 50 sectors and subsectors (i.e., those called out on Fig. 6.1), yet we found that in the end, under the assumptions we have made—which make no special concessions to possible new AI trends—if there is indeed going to be a self-sustainable and commercially viable human outpost on the Moon, it all more or less comes down to one thing—people. Both governmental and commercial. That's what drives the revenue numbers. Maybe that should not be so surprising. We are creating and populating a new world. People living and working on the Moon who will all be mutually dependent. And they will speak different languages, just like here on Earth. It makes sense. They will need homes, and be working in mining and manufacturing. Unless, they are tourists just along for the experience.

This is the material which must be further developed over subsequent years, as better data becomes available. New reviewers can challenge the assumptions that were made. You now have insight into which assumptions were most critical, and which assumptions were subject to the widest range of potential values, and hence uncertainty. You now know where critical market research is needed. You know where we need to obtain some early results from lunar surface experiments in order to validate, or amend, the values used. So, hopefully now you have a basis for considering investment decisions. And you have some ROM perspective for budgetary planning in the national space agencies. If you are a potential entrepreneur in the lunar commerce sector, you now have an angle on the likely ROM revenue values for your chosen sector, and the data elsewhere in this book regarding potential competitors, and thus some idea of your potential revenue range in the Mature Phase. Overall, it seems that, under our stated assumptions, the biggest driver of demand across all sectors comes back to the anticipated number of people on the surface, or in lunar orbit. Of course, it bears repetition that we do not know exactly when that Mature Phase will begin. Seems we ducked that timing question with the way we structured our approach for this Version 1 of the LCP. But you kind of get a gut feel, from viewing the state of readiness, or not, of the various sectors that we have been exploring, that it will not be happening anytime soon.

You will need to look back into the previous sections of this book to understand what the likely competitive scene will be for each of these market sectors (hint: look under "Suppliers"), and the likely impediments to achievement, before deciding whether to become an active investor, or to contribute products or services to some part of the supply chains of these sectors. Also, it is important to review how much, and how likely, will be the necessary government commitments to making their contributions to the creation of the lunar economy.

How good is this work? We have tried our best, but it is nevertheless worth looking, as a reasonability check, at this lunar commerce potential alongside the revenues achieved annually within the traditional existing commercial space business (which largely consists of making launchers and satellites for Earth-orbit purposes). We note that currently the traditional annual commercial space business revenues are about $350 B, split between launchers, satellites and ground equipment. So, we are here developing a lunar commerce marketplace which would provide additionally about one tenth to existing commercial space revenues. Not unreasonable. Also, in addition, we note that space tourism markets in the near-Earth vicinity are currently assumed to be worth about a further $2 B. So, it seems that we are in the right general ballpark, bearing in mind that lunar tourism forecasts underpin much of the identified lunar commerce $ 31 B/year eventual revenue opportunity.

We have used the single set of data represented by the Lunar Commerce Portfolio to drive the values in this book. It would be better perhaps if we had some alternative sources for comparison, so we could tell when our assumptions stack up well against the findings of others, who have used different approaches. So, we have identified some other prior published estimates for lunar revenue potential, but in general these references do not contain a full account of sources, assumptions and models—so a full comparison has not proven possible. Moreover, they are not even trying to compute the same thing. Some are cumulative, some are annual revenues, and each of the sources assumes a different amalgamation and grouping of market segments, and different time periods. We nevertheless provide here those summary findings for reference (PWC, 2021; NSR, 2021; and Citi, 2022). Again, we seem to be in the ballpark as a general reasonability check, with our $31 B/ year figure, given the wide range of uncertainty around the value.

April 2021	NSR Forecasts	$42 B (cumulative to 2030)
November 2021	PWC Forecasts	$170 B (cumulative to 2040)
May 2022	Citibank	$100 B/year (but includes other sectors, e.g., space solar power)

In summary, for the NSR material, there is a charge of either $4795 for access to the "Standard" version, and $8795 for the "Enterprise" version of the report, and it is prohibited to redistribute the material and results beyond the published public domain totals. So, I hope you forgive me if I did not buy one. It is difficult to parse the contributing submarkets from the public domain material, although the overall total is more or less in line with our LCP figure of $31 B, although in the NSR case, these revenues are anticipated to be derived cumulatively by the end of what we are calling the Early Phase. In the case of the PWC forecasts, it appears that the $170 B total consists of about $100 B for transportation, about $60 B for space resources, and $10 B for data. The results, however, are expressed as cumulated revenues over the period to 2040, which explains why they are five times greater than the annualized revenues we have derived. For the Citibank report, the methodology is not fully explained, but it does depend on computing a series of extrapolations using compound growth rates. The report contains a series of interviews with prominent players in the industry, but most of them are involved with the existing government and satellite space businesses, which gives an indication of the source of the driving assumptions. The study posits a growth in the overall space economy from a stated current value of $350 B (2020) to about

$1 Trillion (2040), and of this, roughly $100 B/year is identified as "new applications and industries," a category which, however, includes space solar power, which we have not treated as a lunar commerce business. But, within that "new applications and industries" category, specifically $12 B/year is allocated for Moon/asteroid mining, $8 B/year for space tourism (all types), and $14 B/year for microgravity R&D and Construction. Therefore, there is about $34 B/year for generalized lunar-related business, as best we can discern—not so far from our own results.

So, there it is, the result from crunching all our assumptions is about $31 B/year in lunar business revenues, and this figure is not dramatically out of line with the findings of others. And because all of our assumptions in the Lunar Commerce Portfolio have been explicitly and transparently documented, it I believe provides a solid base for future updating, as better and more recent data becomes available. We look in the next chapter to understand the inherent uncertainties in any forecasts today of a lunar economy.

References

Bocconi. (2022). SDABOCCONI.it/en/news/22/11/a-new-partnership-between-mva-and-SDA-Bocconi-school-of-management

Citi. (2022). *Space—The dawn of a new age*. www.citi.com/citigps

Gautel, G. (2023). *Technical presentation on the lunar commerce portfolio report*. UNCOPUOS.

LCUG. (2022). Lunar Commerce User Group. https://Lunarcug.com

NSR. (2021). *Moon market analysis*. https://www.nsr.com/research/moon-markets-analysis/

PWC. (2021). Space-economy.esa.int/article/119/pwcs-lunar-market-assessment-market-trends-and-challenges-in-the-development-of-a-lunar-economy

12

Risks and Uncertainty

How meaningful is the $31 B/year ROM value of potential commercial lunar revenues? Can we rely on it? Let's add a dose of reality. Our forecasts are subject to multiple layers of uncertainty. As we have made clear, and explicitly documented, these forecasts are a result of a modeling exercise using a series of key assumptions. Any change to those assumptions and the revenue forecasts will change accordingly. We even noted where we had some missing subsector revenue calculation data at this stage (examples included storage, low-g and high-vacuum manufacture for terrestrial uses, and mining of PGMs for Earth use). As was the case on Earth, as markets developed due to early space exploration efforts in the sixties and seventies which were totally unexpected (think GPS-enabling, think personal microcomputers), the same will probably be the case with lunar commerce. There may be markets which we have not considered here, and which could even eventually emerge as the dominant sectors, and in this work to date we assume zero revenue from such potential sources. But at least we have a slot for them in Sector 11. And what about the potential impact of the newly developing AI field on the lunar economy? Any significant move away from the human-tended base-setup we have described toward an AI/robotic lunar operation would certainly undermine, or require a reassessment of, many of the assumptions behind this work, which are premised on the need to support a growing number of human personnel on the Moon. But we do now have a beginning, representing our collective knowledge, and its limitations, in November 2022.

Two main types of assumption will affect all markets. The first type is the list of common driving assumptions included in Chap. 6. In the table in that chapter, we underlined the enormous range of potential values for each of the driving assumptions, due to the presently unknown factors capable of

© The Author(s), under exclusive license to Springer Nature Switzerland AG 2024
D. Webber, *Lunar Commerce*, https://doi.org/10.1007/978-3-031-53421-8_12

influencing these key values. At present, there is a factor of anything from 3 to 30 in setting values even for our driving assumptions. Much of this is due to the uncertainty in lunar space tourism demand. We have included all the associated assumptions in Appendix B, so that improvements in this important area can easily be made in future. However, at this point, it should be restated that these lunar tourism markets were estimated, in the absence of some good statistically valid market survey work, using very conservative assumptions (it was assumed, e.g., that prices for Moon trips would be as high as $150 M for lunar orbit, and $750 M for lunar surface stays; it assumed that only people who could afford such prices without blinking would be candidates—the ticket price would have to be less than 10% of their net worth, and furthermore that only 5% of such people would want to do it). In fact, the cut-off point was having a net worth of at least $1.5 B for lunar orbit, and $7.5 B for lunar surface tourism. And yes, we checked, there are enough of such people around. We have to hope that they will enjoy each other's company. Also critical to lunar commerce market expansion is the success and availability of the SpaceX Starship, with associated low transportation prices for people and cargo. Fingers crossed on that one, because it is the key technological building block to making any of this happen. SpaceX takes enormous risks every time they conduct test firings. Incidentally, for this Version 1 of the Lunar Commerce Portfolio, we decided *a priori* that we would, for simplicity, *exclude* all markets that would result on the Moon from future ongoing Mars missions. This is not a small matter, if we are thinking of the Moon as an orbiting gas station exactly to perform that function, for craft heading outward beyond the Moon. We shall certainly need to include it for future Versions of the LCP.

Then there are the vast number of alternative assumptions which go to describe each of the possible scenarios. They are also listed in Chap. 6, and in Appendix B. As a general statement, the level of risks and uncertainty are liable to be greater in the Mature Phase than in the Early Phase. Because at this stage of lunar commerce development, so little is known, we have simply set all values based upon a consensus view of the international team of volunteer analysts working within the Moon Village Association. The model has been made available for free to the community, so we can all manipulate the assumptions one at a time to discover the resulting impact. For our purposes here, we simply report on the uncertainties that result from switching from one scenario to another. We compare the results of Scenario Alpha—the least optimistic—with Scenario Delta—the most optimistic (see Fig. 11.1). The revenue projection varies by a factor of almost 8 (i.e., $4 B vs. $31 B) in the expected outcome, depending on the changes in this set of external factors.

And in addition to these overriding assumptions, which affect all market sectors, there are sector-specific assumptions (e.g., price levels for services) which just affect the values for each respective sector in isolation. So, there is, we conclude, a lot of uncertainty. To put it mildly.

But, at least with this technique we have the ability to incrementally over time improve the confidence level of our stated assumptions, and this will in turn result in less uncertainty. How can we do this? Well, we have identified the assumptions which carry the most uncertainty, and so we must work at understanding these interactions better. For instance, a major key assumption involves lunar surface space tourists, their numbers, and what they will be willing to pay for the experience. It would therefore be helpful if someone could put in the public domain the results of a statistically valid survey of very rich individuals from all over the world, which would lead to more confidence in the numbers and price levels. Other assumptions involve knowing to what degree we shall find water on the Moon, and how easy it will be to mine it and render it useful for drinking, and for other purposes, such as making rocket fuel. The only way to reduce the uncertainty on these particular assumptions is to encourage more experimentation with precursor landers on the Moon, and publish the findings. Meanwhile, we must accept that the $31 B/year ROM should really be seen as representing a range of from maybe round about $5 B/year to maybe up to around $100 B/year. This is the reality check. And as pointed out, much of the uncertainty is due to not knowing whether a successful lunar tourism market will develop, in what timescale, and at what prices. But at least we have a good structure, and a good model, in order to monitor and reflect the future changing knowledge about the commercial lunar markets. We also need to get a better handle on the "Earth-scarcity" markets—Platinum Group Metals, rare earths, and maybe He3—and include their revenue potential in the next Version of the LCP, because they, also, are part of the prime rationale for going back to the Moon for resources, and at present, we have not been able to provide revenues.

For new firms to secure capital funding, and for insurers to be willing to underwrite the project, there needs to be much less uncertainty in the revenue projections. They also need to have a better idea of realistic timeframes. What we have found in this study is that the true commercial lunar economy will likely not happen in the six remaining years of this decade. Yet to obtain VC funding, normally the providers of finance look to see positive returns on investment in only a few years. Such funders, it seems, will only have a chance of meeting these criteria if they are working to support the first few market categories, which as we have seen will be the ones that for sure are receiving government contracts. We are talking about sectors 1–5. It is therefore unlikely that venture capitalists will provide funds for sectors 6 onward until they have seen the success of the Early Phase between now and 2030. Moving beyond that point, and requiring the funds for true commercial ventures, will not be possible before the overall uncertainties in the size and timing of the lunar commerce marketplace will have been reduced by the passage of time, the availability of new technology experiments, the results of carefully constructed and managed market research studies, and more statements of commitment by potential providers. With this work, and in particular as it is improved by the indicated further work by those anticipating operating in this business sector, leading to a reduction in forecasted revenue uncertainties, we shall have effectively established the imperative. Only by reducing the uncertainties, in the ways described above, can we reduce the potential risks for investors.

Incidentally, a significant part of the uncertainty, as represented by the scenario assumptions, is a consequence of the unknowns in the way the regulatory bodies in the world and nationally will behave in addressing some of the implicit questions in these scenario choices. We come back to these matters in Part IV.

Well, we delivered on our promise, I think you will agree. You now have the numbers, and all the assumptions behind them. Or at least you have a first-cut at the numbers. We need more work. We may recall in Douglas Adams' wonderful book "The Hitchhikers' Guide to the Galaxy" (Adams, 1979) that the ultimate answer turned out to be 42. Something that Elon Musk celebrated when he launched his Tesla Model S roadster to Mars and beyond in 2018—with a copy of the book in the glovebox (Musk, 2018). With our first efforts at quantifying the lunar economy, the magic number turns out to be 31. Give or take a bit.

References

Adams, D. (1979). *The Hitchhiker's guide to the galaxy*. Pan Books.
Musk, E. (2018). *Musk uses his car as dummy payload*. https://en.wikipedia.org/wiki/elon_musk%27s_Tesla_Roadster

13

Back on Earth

We have gone as far as the current data takes us regarding the scale, timing, and risks of the potential new lunar commerce businesses. Some of them would be offshoots of already existing terrestrial companies; others would be entirely new entrepreneurial operations. Some would require knowledge of operating in space; others not so much. They will potentially be from a wide array of countries on Earth. There will have to have been international agreements set in place to enable the activities to take place (to be discussed in Part IV). There will be a mixture of governmental and true commercial customers for the products and services. The sectors we have described would have enabled a sustainable way for folks to live and work on the Moon, and not be dependent on deliveries from Earth. They (the lunar inhabitants and visitors) would be effectively autonomous. This would be the first time that we have imagined an "off-Earth" commercial ecosystem, and so we should try to foresee how that situation might impact life, and business, on Earth. What might be the implications? What might be the impact on traditional economic activities on Earth of having this lunar economy?

One interesting question would be who would tax them? How feasible would it be to attempt to keep track of the various national terrestrial origins of the supply chains going to make each of the lunar businesses? Will the Moon become "the ultimate off-shore" location for tax purposes? It would certainly prove to be a test of the extent of the presumed, and even defined, autonomy of those conducting lunar commerce. Another question would involve currency. Would there emerge a lunar currency for purposes of transactions between lunar businesses, and indeed for purposes of trade with terrestrial entities? And how would the value of those currency holdings be arbitrated in terrestrial foreign exchange markets?

D. Webber, *Lunar Commerce*, https://doi.org/10.1007/978-3-031-53421-8_13

Even though we have defined the lunar commerce "Mature Phase" as being one of self-sufficiency, in that the necessities of life will be provided from lunar resources, that does not mean that there would not continue to be *trading* with the Earth. Lunar inhabitants will continue to want to purchase nonsurvival and more luxury items for delivery from Earth. Also, we have seen that some of the lunar businesses intend to export their product back to Earth. This will make most sense when there is a need on Earth due to rarity. This might include, for example, bulk shipments of lunar-sourced PGMs. Clearly, the pricing of such lunar exports will have to be arrived at to avoid major disruption to the economics of traditional Earth-based extraction industries. Rarity in itself gives rise to high prices. So, we might assume that Earth's commercial actors in these areas of scarcity will become involved in the lunar commercial activities, so that they will be able to better manage the economic impacts.

Also, there will be an aspect of the developing trading scene that is best viewed as *competition* between the products produced on the Moon and those produced on Earth. Clearly, the lunar-sourced products would have a major disadvantage, in that the transportation costs from Moon to Earth must be added to prices offered to terrestrial consumers. So, this will not emerge as a problem for terrestrial providers in the normal way. It will only make sense if lunar-based manufacturers are able to take advantage of the properties of the lunar environment (high vacuum, low gravity, etc.) to make products much better than could be achieved on Earth. Only time will tell if this will be possible (when costs of delivery from Moon to Earth have to be added). We might even imagine, at some distant future time, that terrestrial manufacturers might levy their governments for import taxes on lunar products.

In summary, how will lunar enterprise differ from Earth's commercial activities? There are differences which you should take into account if you are new to space investing. For instance, there is of course no track record to rely upon—at present nobody has ever done this before. Which means that there are inevitably a whole lot of "unknown unknowns" to be added to the list of "known unknowns." With regard to the "known unknowns," there is plenty of information on subsectors such as the various envisioned subelements of the lunar mining process (Lewis, 2015). Some early work on 3D manufacturing in vacuum has been conducted on the ISS, and lessons learned. But clearly, to date all bulk work on construction has had to rely on various lunar regolith simulants. And none of them can completely replicate what it will really be like on the surface of the Moon.

Another key factor is that, in general, space is highly regulated, but the regulation cannot keep pace with the technological changes (Anderson, 2023). So, there are fuzzy boundaries—which can for sure provide an opportunity for disruptive startups, but on the other hand, this can create difficulties in planning. In fact, planning in general needs to emphasize flexibility for those contemplating being part of a lunar business enterprise, even when a lunar developer is a subsidiary of an existing terrestrial business, where perhaps planning has been conducted in the past on more stable foundations.

We always know in creating businesses on Earth, that listening to your customer is of the highest priority. So much so that companies spend large amounts of money conducting market research to establish in some detail what the customer is seeking. This fuels future developments and also provides valuable input for the service and maintenance functions. This will be no different on the Moon. However, it might prove to be more difficult to figure out literally who is your customer. For example, for startups, government agencies can be customers, competitors, or both. This whole concept of the Lunar Commerce Portfolio began (Webber, 2020) with trying to be clear on the definitions of even what "commerce" means, when a government entity can be both a customer, a regulator, and a provider of funds.

In the context of knowing your customer, of particular importance on the Moon, under our assumptions, will be the inherent dependency on others. Everyone on the Moon will be, to a greater or lesser degree, dependent on everyone else. Everybody is everybody else's customer. We have seen that through our work, as summarized in our market sector summary charts. Therefore, all companies will have some vested interest in seeing their neighbors succeed—even when they might be competitors in some degree. So, this is new. Some of the first "bases" established by Europeans in the New World of the Americas did not succeed, and their inhabitants disappeared without a trace. So, some kind of community involvement will be a requirement of these first "settlers" on the Moon, and there will likely be a counterpart on Earth of the terrestrial counterparts, or owners, of these lunar business ventures, which can work together to interface with governmental entities on Earth and can in general help make the lunar outpost a commercial success. Their shareholders will expect nothing less. We hope that the new organization the Lunar Commerce User Group will be able to make progress with that shared understanding, and thereby be a forerunner to the eventual lunar commerce partnership.

References

Anderson, C. (2023). *The space economy*. Wiley.

Lewis, J. (2015). *Asteroid mining 101* (pp. 112, 113). Deep Space Industries.

Webber, D. (2020) Lunar commerce: A question of semantics? *The Space Review.* 16 Nov 2020.

Part IV

Framing the Way

Having established some ranges of potential revenues from future lunar businesses in Section III, and the dependency on, amongst other things, the outcome of international and national directives, this section explores what must now be done to frame the way forward in terms of policy and regulatory provisions.

Also discussed in this section are some of the ethical issues, and military and political factors, which may affect regulation, and some steps being currently undertaken to guide progress.

And, finally, the conclusions are presented, which sum up the findings of this first version of the Lunar Commerce Portfolio, and which suggest the next steps.

"There is a tide in the affairs of men,
Which, taken at the flood, leads on to fortune,
Omitted, all the voyage of their life
Is bound in shallows and in miseries"
– Shakespeare, Julius Caesar

14

Policy

Why are we spending so much to go into space? Why do we need astronauts at all? These have been the often-expressed views of the general public (i.e., us), especially in the USA, and especially after the success of the Apollo program. Both in the USA and elsewhere in the world, there has been a long succession of space policy initiatives over the decades since the beginning of the space age back in 1957. But the overall impression, at least to the US man-or-woman-in-the-street, is that nowadays our reasons for being in space lack clarity. At the beginning, it was clear. The USA had to do it because the Soviets were doing it. And maybe satellites would in some way be used as weapons. Some of this subsequent space policy has resulted in regulation, both global and national. And those regulatory aspects will be discussed in Chap. 15. In other parts of the world, however, there is emerging a new focus and expectation for space and its potential in terms of benefits for us on Earth.

In some ways, the development of space policy can reflect on ethical matters. Sometimes, it involves military or other strategic considerations. Sometimes, it is a vehicle for procurement discussions. Generally speaking, though, policy statements are an expression of the space "vision" at different points in the historical development of the technologies. Nowadays, ethical issues include matters such as whether any country, or national entity, has any right at all to perform certain proposed space activities. Such as mining asteroids, for example. Or, whether the Moon might be considered as a commercial resource, or alternatively should be treated exclusively as a scientific protected site. And if there should be protected historical sites on the Moon, for instance, what would such protections entail? And who would determine which such sites could meet the international criteria for protection? What kinds of environmental protections (what does that even *mean* on the Moon?)

D. Webber, *Lunar Commerce*, https://doi.org/10.1007/978-3-031-53421-8_14

might be needed for the Moon? The international GEGSLA group, initiated by the MVA, has been working to address many of these issues (GEGSLA, 2023). GEGSLA has a long list of matters to consider, which at present includes such items as, e.g., lunar registry of objects and activities, boundaries and safety zones, heritage sites, prioritization of lunar use activities, sharing mechanisms, and experimental zones.

There can also be cultural considerations, especially when taking an international perspective. The Moon has a significance in many ways, and in many different fields—such as in religion and the arts. There is for instance some glorious operatic music inspired by the Moon—consider "Song to the Moon" in Dvorak's "Rusalka." Native cultures in the Americas and elsewhere revere the Moon. They have rituals, poetry and folksong traditions. When we look at the Moon at night, do we see a man's face, or do we see a rabbit? Would lunar mining maybe interfere with that?

Let's check out the policy initiatives from around the world which could impact the potential for lunar commerce. We are going to list below some of the national space policy perspectives from around the world, from existing players, so that we can see how national actors might help, or hinder, the attempts to commercialize the Moon. But remember that the coming "return to the Moon" era has also become a point of focus for countries who have *not* previously had a space program. They want to become involved and find some place in the value chain where they can contribute. This material is only up to date as far as August 2023—and things are moving fast.

USA

After some very clear visionary space policy statements and rationales at the outset of the space program, notably by President Kennedy (Kennedy, 1961), and despite several attempts by subsequent US presidents to re-engage the public in grand initiatives, they have in general not succeeded in igniting the public interest, and in the USA, funding for NASA's space activities has dropped in percentage terms to one tenth of its levels back in the sixties. Meanwhile, in other parts of the world, there has been more rhetorical progress—although as we have seen, the US budgetary support for its space program by far still outweighs that of any other country. Some would claim that, after the initial success of the Apollo era, the USA had lost its way in space. Certainly, there have been many very successful robotic scientific and exploratory missions throughout the solar system during the last half-century, but there has been a lack of clarity, in particular concerning the rationale for having human crews in space.

This began to change with statements by the Presidential adviser John Marburger in 2006 (Marburger, 2006). He introduced the idea that the inhabitants of Earth could and probably should view the resources of the solar system as part of our economic portfolio. More gradually, language about "settlement" has begun to be included in US space strategy documents, and this at last makes sense of why humans need to be in space.

Perhaps, the main US policy statements over the years have been the following.

1958	Eisenhower—Created NASA (Eisenhower, 1958)
1961	Kennedy—Moon landing commitment (Kennedy, 1961)
1967	Johnson—Proposed Outer Space Treaty. Intelsat created in 1971 (Johnson, 1967)
1972	Nixon—Ended Apollo—initiated Space Shuttle. Apollo/Soyuz Test Project (Nixon, 1972).
1981	Reagan—Int Space Station and Commercialization policy (Reagan, 1988)
1989	GHW Bush—Space Exploration Initiative (SEI); Permanent settlement on the Moon (Bush41, 1989)
1996	Clinton—National Space Policy. "Enhance knowledge" and "Security of the US" (Clinton, 1996)
2004	Bush (43)—Vision for Space Exploration (VSE). Human return to Moon by 2020. Constellation. National Space Policy (Security and Private enterprise). (Bush43, 2004)
2009	Obama—Canceled Constellation, instituted SLS heavy lift. Introduced US Commercial Space Launch Competitiveness Act.
2015	Obama—Signed SPACE Act "Spurring Private Aerospace Competitiveness and Entrepreneurship," also known as Commercial Space Launch Act of 2015, which allows a US citizen to possess and sell space resources obtained from the Moon or asteroids.
2017	Trump—Back to Moon by 2024, with commercial and international partners. Establishment of US Space Force (Trump, 2017).
2020	Artemis Accords (Artemis, 2023, and the Appendix).
2021	Biden—US Space Priorities Framework—understand climate change (Biden, 2021).
2023	NASA Budget request problems (NASA Budget, 2023)

Meanwhile, other space agencies have expressed their own visions of space exploration and development, and we note here how they may lead to, or impact, lunar activities.

Europe/ESA

Europe, through its space agency ESA, has, over the years, sought to have its own access to space via its own (Ariane) rockets, and fleets of Earth observation, communications, and navigation satellites (Galileo). Key European space initiatives have been the following.

1961	Creation of ELDO (European Launcher Development Organization)
1961	Creation of ESRO (European Space Research Organization). First satellite launched (ESRO II) in 1968.
1975	Creation of ESA (European Space Agency—combination of ELDO and ESRO)
1977	Creation of EUTELSAT (European Telecommunications Satellite Organization—first launch 1983)
1979	First launch of Ariane (Arianespace formed 1980), from Guiana space center
1979	Creation of INMARSAT (International Maritime Satellite Organization—based in London).
1983	SPACELAB launched, and Ulf Merbold first ESA astronaut. Support for ISS.
1986	Creation of EUMETSAT (European Meteorological Satellite Organization)
1986	Giotto comet probe
2003	SMART-1 probe to Moon. Achieved lunar orbit, and hard landing in 2006
2013	Orion's Service Module (ESM) approved for Artemis Moon missions 1 thru 5 (allowing 3 European astronauts to visit lunar orbit at the Lunar Gateway).
2021	Matosinhos Manifesto—Moon Ice Sample Return, Green initiatives, Security of Space Assets (MM, 2021).

ESA's HLAG (High Level Advisory Group) produced in 2022 the report "Space Revolution Report of the High-Level Advisory Group on Human and Robotic Space Exploration for Europe," in which they recommend that ESA should land its astronauts on the surface of the Moon by 2033, with its own rockets, spacecraft, and lander.

Russia

Russia was, and continues to be, an important country in the development of space. In fact, for many years at the beginning of the space age, Russia (or more correctly, the former Soviet Union) was leading technologically (e.g., first satellite Sputnik 1, 1957, first man in space, Gagarin, 1961). Russia was also the first to reach the Moon (Luna-2 in 1959), to image the far side of the Moon (Luna 3 in 1959), and the first to operate a rover on the Moon (Lunokhod in 1970). However, they have not succeeded in landing humans on the lunar surface. Russia has placed probes on Mars and Venus, and the Soyuz rocket and spacecraft have continued to be a mainstay of the International Space Station, taking astronauts from many countries, and indeed some space tourists, to their orbital destinations. However, in the USA, alternative means have been developed, largely as a consequence of the "New Space" movement. So, Russia is losing a vital export market, both for the spacecraft, and the RD-180 engines. With the collapse of the Soviet Union (1991), there was considerable confusion within Russian space circles, but eventually Roscosmos emerged in 2015 as the effective manager of the Russian space program. The author, for example, Webber (2017), was at that time trying to negotiate for a Proton launch for the Inmarsat organization, and could not easily determine who in post-Soviet Russia was the "owner" with whom to negotiate, or even find anyone who knew its price in hard currency. Cooperation with Europe had been an important part of the Russian program going forward, but geopolitical events (Ukraine invasion, and sanctions) have made continuation of this unlikely. In an attempt to arrest the decline in Russian space activities, the main emphasis has been to make the Vostochny Cosmodrome operational (in order to be independent of the former Soviet launch facilities in Baikonur, which now reside in the independent nation of Kazakhstan), to update the launcher fleet (to Angara rockets), and maintaining satellite constellations (for communications, navigation, and Earth observation). It is considering developing a new Russian orbital station for launch at the demise of ISS (after 2030). The country has been extending its partnerships to South Korea, Israel, Japan, Brazil, China, and India. Russia has recently begun discussions with China concerning having a joint lunar base, the International Lunar Research Station, established (Russia/China, 2021).

China

China's involvement in space goes back to 1970, when it launched its first satellite Dong Fang Hong, and the country has displayed throughout a highly competent and systematic approach to the exploration of space. China has Earth-resource satellites, a navigation satellite system (Beidou), and communications satellites. Even our Fig. 1.1 is a photograph obtained by a Chinese spacecraft way beyond the Moon, looking back toward Earth. At the time of writing, China is still only the third country to have launched astronauts (called Taikonauts) into space, starting in 2003 (Yang Liwei was first). They have their own spacecraft, called Shenzhou, based somewhat on the Russian Soyuz. China has a stable of rockets (called Long March) with eventually a good track record, it has its own space station (Tiangong), it has delivered payloads to the Moon and Mars, and the country is beginning to plan for a human lunar landing, and a possible joint Russia/China lunar base (Russia/China, 2021). China has tried to commercialize its rocket-launch services, but with limited success, largely due to a sanction regime. Formally, the China National Space Administration (CNSA) has announced their long-term goals to include establishing a crewed space station, crewed missions to the Moon, establishing a crewed lunar base (the International Lunar Research Station—ILRS—jointly with Russia), a robotic mission to Mars and industrial development of cis-lunar space (China, 2022).

With specific reference to the Moon, China's activities have proceeded as follows;

2004 CNSA announces uncrewed Moon exploration project.
2007 Chang'e 1 first Chinese lunar orbiter.
2013 Chang'e 3 soft-landed on Moon, and deployed the Yutu rover.
2019 Queqiao reached a lunar mission orbit as a lunar relay satellite.
2019 Chang'e 4 soft landed on Moon's far-side, and deployed Yutu-2 rover.
2020 Chang'e 5 returned lunar samples to Earth (only the third country to do so).

Planned future missions include

Chang'e 6 to collect and return lunar far side samples
Chang'e 7 target is lunar polar region searching for water ice deposits.

Chinese astronauts to land on Moon by 2030.

China is also proceeding to build a launch facility in Djibouti in Africa, which is a state that is nonparty to the treaties governing outer space behavior. A contract was signed in January 2023 to begin work, and it is not clear what impact this development might have on previous norms of behavior.

Japan

Japanese involvement in space arguably started in 1955, with a first rocket experiment. Since then, there have been some singular achievements, such as the 2005 Hayabusa mission to the asteroid Itokawa, during which samples were taken, and returned to Earth in 2010. The Japanese space agency JAXA was formed in 2003, and by 2007 had managed to successfully place Kaguya/Selene into lunar orbit (it eventually hard-landed on the Moon in 2009). Japan also launched a Venus orbiter, and built a module called Kibo for the International Space Station (2009). Its astronauts have visited the ISS since that time (launched either on the US Space Shuttle or the Russian Soyuz). In the domain of private spaceflight, we can note that the journalist Akiyama in 1990 became the first effective space tourist flying to space station Mir by Soyuz (although his ticket was paid for by his newspaper employer—Dennis Tito of the USA was the first space tourist to buy a ticket with his own funds). And, furthermore, the Google Lunar XPRIZE included the Japanese Team Hakuto, who were building a lunar rover in an attempt to win a prize. After the GLXP ended, Hakuto joined with iSpace and in 2022 entered lunar orbit with Hakuto-R, as part of NASA's CLPS initiative. By 2023, they were ready to attempt the soft landing, but did not succeed. iSpace/Hakuto is preparing a backup lander/rover. Japan is a participant in the Artemis Program. The main focus of Japan's space policy is (1) ensuring space security, (2) contributing to disaster management, (3) creation of knowledge through space science and exploration, and (4) realizing economic growth and innovation. In 2021, Japan enacted the "Act on the Promotion of Business Activities for the Exploration and Development of Space Resources." Specifically related to lunar exploration, Japan intends to be contributing to Artemis with habitation technology, resupply, lunar mobility, and lunar nav/comms.

India

India has had a very long and successful involvement with using space. The national space agency ISRO was formed in 1969, launched its first satellite in 1975 (using a Soviet launch vehicle), and then launched its own satellite on its own launcher in 1980. Most of the many launches over the years have been for communications (INSAT), navigation (IRNSS/Gagan), and Earth observation (IRS) purposes. Telemedicine was a key part of the early comms satellites. However, there have also been space exploration missions, including the 2014 Mangalyaan which went into Mars orbit. There are plans for a Venus orbiter Shukrayaan-1 and a Jupiter/Asteroid mission. India has yet to launch a human crew into space, but has been preparing the Gaganyaan spacecraft for the purpose, to be followed by an indigenous space station (India, 2023). The Indian human space program is imminent.

With regard to the Moon, there was an Indian Team (Team Indus) competing within the Google Lunar XPRIZE, and which won some interim prize money. The team of international GLXP judges was very impressed with the commitment of all the junior engineers and their work product. When the GLXP closed, Team Indus was awarded funds under the NASA CLPS program (but that prize was withdrawn when NASA insisted the craft be made in the USA, not in India). Nevertheless, India continued to try to land a robot spacecraft to operate on the lunar surface, and Chandrayaan-3, having been launched in 2023 has succeeded. Chandrayaan-1 was launched into lunar orbit in 2008, Chandrayaan-2 hard-landed in 2019, and Chandrayaan-3 made a successful landing and robotic rover deployment in the polar region on August 23, 2023. We can probably reasonably anticipate that there will eventually be Indian astronauts on the Moon.

The Indian Space Policy of 2023 was formulated and supported "a flourishing commercial presence in space" by "encouraging and promoting greater private sector participation in the entire value chain of the space economy."

Although a signee of the Artemis Accords in June 2023, India announced in August 2023 that it is working with other BRICS nations (i.e., Brazil, Russia, India, China, and South Africa) on space partnership plans, which may in some respect be considered to be a counter to the US-led Artemis Accords. At the same time, it was announced that six other nations (Argentina, Egypt, Ethiopia, Iran, Saudi Arabia, and UAE) were joining BRICS starting in January 2024

UAE

The Middle East is emerging as a vibrant sector in space development. The UAE has been particularly prominent. It manages satellite fleets (Yahsat and Thuraya), and has an astronaut corps with astronauts going to the ISS. The UAE Space Agency was established in 2014, and the sector is supported by a $1 B fund. The programs are ambitious, as described in the 2016 space policy document (UAE, 2016), and include "manned and unmanned exploration missions to enable human colonization of space," and "exploration, mining, extraction and utilization of resources in space." Further text from that policy document refers to the need to promote creative entrepreneurship and commercial space projects, with cited specific examples: "space projects related to space manufacturing by robots, three-dimensional printing, commercial space flights (orbital and suborbital),…the exploration and exploitation of resources in space….that may create a global revolution in the field of exploration, the utilization of space and its resources, and the spinoff of space technologies." This is not at all empty rhetoric. They have an Act on their books allowing for ownership of space resources. In 2020, the UAE successfully sent a probe called Hope to Mars, and in 2022 the Rashid rover was launched to the Moon on board the iSpace Hakuto-R Mission 1 lander, which unfortunately hard-landed on reaching the Moon. There are plans to engage the space tourism sector, originally involving Virgin Galactic, and the UAE Space Agency has signed the Artemis Accords.

Saudi Arabia

Also representing the Middle East, Saudi Arabia has been active in the space domain for some time. Arabsat has operated a satellite telecommunications business since 1976. In 2018, the Saudi Space Commission was established, and is supported by a $2 B fund. Two Saudi astronauts are scheduled to go to the ISS this year (2023), with help from the Axiom organization. Saudi Arabia has signed the Artemis Accords.

Israel

The Israel Space Agency was founded in 1983, and has the capability to build and launch its own satellites, specializing in communications (Amos) and imaging (Eros) satellites for LEO and GEO use. They have particular expertise in microsatellites. One Israeli team (Team SpaceIL) competed in the Google Lunar XPRIZE competition, and they built the Beresheet lunar lander and launched it to the Moon in 2019. It arrived in lunar orbit and sent back images (receiving as a consequence a small award from the GLXP, although they were too late for the competition proper, which had ended in 2018), but did not succeed with the soft landing. A follow-up Beresheet 2 spacecraft is being prepared to try again in 2025.

Luxembourg

This small resource-constrained country has nevertheless managed to be a significant player in the utilization of space, mainly due to its commercially directed policies, including provision of grants and investments and supportive legal frameworks. As such, it might be seen as a model for other nonspace-faring countries who intend to become part of the upcoming space development activities. Initially focused on satellite communications, Luxembourg founded SES in 1985, and they became a successful satellite services provider starting with their first in orbit Astra satellite operations in 1988. Luxembourg is very much involved in international space ventures, and is an ESA member state (since 2005), and a member of UN's COPUOS (since 2014), and more recently has established partnerships with Japan, China, and the UAE. In recent years, the Grand Duchy has changed focus, and been quick to recognize, and seize on the potential of, space resources. This new focus began with a 2016 initiative (Luxembourg, 2018), followed by the passing of a 2017 law which allows companies to own resources extracted from celestial bodies. Luxembourg established its space agency in 2018, and has subsequently established a world center of excellence related to space resources. As a direct consequence of the Luxembourg government's commercially supportive regime for space developments, several entrepreneurial ventures have moved their operation to the Grand Duchy, including in 2017 the Moon-mining Japanese startup iSpace, which has been pursuing lunar landing and exploration activities within NASA's CLPS program.

Canada

Canada's involvement in space started with Alouette 1 launched in 1962. An initial focus on communications satellites resulted in the creation of Telesat Canada in 1969, with the launch of Anik 1 in 1972. Through Teleglobe (created in 1975), Canada was an active and effective member of the international telecommunications organizations Intelsat and Inmarsat, punching well beyond their relative investment share, owing to their commitment and active involvement. Canada also early on developed a robotics capability which resulted in the Canadarm, built by MDA, and first launched in 1981 on Shuttle Columbia as a major component of the NASA Space Shuttle (and subsequently also on the ISS). This led to the creation of a Canadian astronaut corps, and Canadian astronaut Marc Garneau first flew in 1984. The Canadian Space Agency (CSA) was formed in 1990. Canada has also contributed scientific instruments on a number of interplanetary missions (including to Mars in 2008), and supported the Radarsat Earth-observation missions since 1995 (Canada, 2014). International involvement includes membership of ESA, agreements with Brazil (1965), Japan (1986), Russia (1990), China (1995), and India (1996). Canada has signed the Artemis Accords and is contracted to provide a Canadarm robotic manipulator for the Artemis program's Lunar Orbit Gateway.

European National (UK, France, Germany, Italy)

Although generally members of ESA, which absorbs the greater part of their space budgets, some European countries also operate their own national space agencies. They in general have not shown much interest in pursuing the return to the Moon activities, at least as independent countries. The UK launched its own satellite (X3 Prospero) on its own rocket (Black Arrow) in 1971 from Woomera, Australia, before withdrawing from the launcher business. It has mainly focused its efforts on communications satellite manufacture. The country has astronaut experience via the ESA membership and has signed the Artemis Accords. It is somewhat belatedly trying to re-emerge as a provider of rockets and launch facilities, and as a provider of cubesats, and moreover is establishing a UK Space Command. The French space agency is called CNES. France was launching sounding rockets in 1952, and launched its first satellite Asterix in 1965 from the Algerian desert using its Diamant rocket. It operates the Kourou spaceport for ESA in Guiana, very near the equator, and so therefore helpful for launches of communications and broadcast satellites into the Geostationary orbit, and leads the Arianespace launch vehicle business. French astronauts have traveled to the space station via Shuttle and Soyuz. France has also signed the Artemis Accords. The German space agency (DLR) was founded in 1969. German astronauts are members of ESA's astronaut corps, and Germany has focused its ESA contributions on mission control activities. Germany has interests in developing electric propulsion. So far (mid 2023) Germany has not signed the Artemis Accords. Italy's space agency (ASI) was founded in 1988, although its first national satellite was launched in 1967. As a country, it has focused on ground control for communications satellite systems. Its main contribution to ESA has been the Vega rocket, and Italian astronauts are members of the ESA astronaut corps. Italy has signed the Artemis Accords. Bocconi University in Milan signed an MoU with the Moon Village Association to help develop the knowledge base for the Lunar Commerce Portfolio. Despite most having signed the Artemis Accords, these individual European nations, with their long experience in the space business, do not seem in general to be individually actively pursuing interests in the commercialization of the Moon; they prefer to engage via their membership of ESA.

New Zealand

The New Zealand Space Agency (NZSA) has only been in existence since 2016, but has signed the Artemis Accords. The main drive behind the establishment of the agency was the building of a launch site and the 2009 launch of an indigenous rocket by the NZ firm Rocket Lab (now a US company). New Zealand is also interested in supporting space tourism ventures, and the suborbital craft Aurora is being built there by Dawn Aerospace. The country has just released (2023) its first Aerospace Strategy, which runs to 2030, and the sector has received an initial $28 M of funding.

International/UN Perspectives

By its very nature, the realm of space and space exploration requires a global perspective. In fact, it began with the International Geophysical Year which led to the first artificial satellites (Sputnik and Vanguard/Explorer) in 1957. In the next chapter, we'll see what has emerged since the beginning with regard to a global overview of space activities. Furthermore, we'll begin to look at the future of global regulatory agreements, with specific focus on the Moon.

In democratic societies, whatever finds its way into space policy should reflect the will of the people. To determine what that is, is a political process, which can involve meetings, surveys, and letters to the editor, and to local and national representatives. There is a need to weigh and balance alternative perspectives and come up with a negotiated agreed way to progress. For those opposed, for whatever reason, to the commercial development of the Moon, this, then, is their opportunity to express reservations and may contribute to arriving at a compromise. And this same process must take place at both national and international levels, using all available mechanisms. This is the area where the MVA and the GEGSLA volunteers are trying to come up with a consensus approach. This can take some time, depending on political urgencies, and world events, but I am convinced that there is a way to resolve some of the expressed concerns by creating an internationally-accepted limited "experimental zone" where we can begin to safely and carefully assess the validity of our proposals for lunar commerce, an idea which is further developed in the next chapter.

References

Artemis Accords. (2023). NASA.gov/specials/Artemis-accords/index.html

Biden. (2021). *United States space priorities framework.* https://www.theverge.com/2021/12/1/22811737/national-space-council-kamala-harris-framework-prioorities-climate-change

Bush41. (1989). "Space exploration initiative" and Augustine report "Advisory Committee on the future of the US space program".

Bush43. (2004). *The vision for space exploration.* http://www.NASA.gov/pdf/55583main_vision_space_exploration2.pdf

Canada. (2014). *Canada's space policy framework – Launching the next generation.* Canadian Space Agency.

China. (2022). China presents space plans and priorities in new white paper. *Space News*, Jan 28, Jones, A.

Clinton. (1996). National space policy – Presidential decision directive/NSC-49/NSTC-8.

Eisenhower. (1958). *The national aeronautics and space act.* US Congress.

GEGSLA. (2023). https://moonvillageassociation.org/global-expert-group-on-sustainable-lunar-activities-GEGSLA/

India. (2023). Indian space policy 2023 – Government of India.

Johnson. (1967). "Outer space treaty", Treaty on principles governing the activities of states in the exploration and use of outer space, including the moon and other celestial bodies – UN Office for Outer Space Affairs.

Kennedy. (1961 January). *State of the union address.* US Government

Luxembourg. (2018). *A model for space sector growth: A Luxembourg case study.* The Aerospace Corporation.

Marburger. (2006). *2006 Goddard memorial symposium keynote address.* NASA.gov

MM. (2021). The Matosinhos Manifesto: Accelerating the use of space in Europe. Aschbacher, J ESA/C-M/CCCII/Res.1(Final).

NASA Budget. (2023). NASA warns of devastating impacts of potential budget cuts. *Space News*, Foust, J, March 23, 2023.

Nixon. (1972). *Statement by president nixon.* https://history.NASA.gov/stsnixon.htm

Obama. (2009). *President Barack Obama on space exploration in the 21st century.* http://NASA.gov/news/media/trans/obama_KSC_trans.html

Reagan. (1988). *Most Reagan officials back satellite exports to China.* https://www.nytimes.com/1988/09/09/business/most-reagan-officials-back-satellite-exports-to-china.html – Gordon, MR.

Russia/China. (2021). Russia's space policy- the path of decline?. Vidal, F – French Institute of International Relations. And "International Lunar Research Station (ILRS) – Wikipedia entry.

Trump. (2017). Space policy directive 1 – National space policy of the United States of America" http://trumpwhitehouse.archives.gov/wp-content/uploads/2020/12/national-space-policy.pdf (also creation of United States Space Force NDAA FY 2020).

UAE. (2016). UAE government space policy – UAE space agency.

Webber. (2017). *No bucks, no Buck Rogers – Creating the business of commercial space.* Curtis Press.

15

Regulation

There is no doubt about it. We are going to need a great deal of clarifying and enabling legislation, both internationally and domestically, before we can create a full lunar economy. At an international level, it is still not clear whether we should maybe treat the Moon in the same way as Antarctica. That would require a separate treaty. There is already a "Moon Treaty," but none of the key states has signed it, because of differing views on implementation.

Normally, it takes many years for an UN-type body to come up with an international agreement which can be agreed by consensus. Yet, the Artemis program is already under way, and there are no agreed global international agreements on whether it should be permitted for commercial entities to mine on the Moon. It is likely that discussion on this will continue throughout much of the Early Phase.

There are a series of existing instruments of space law, administered by the United Nations Office for Outer Space Affairs, based in Vienna (International Space Law, 2017). This body also hosts the United Nations Committee on the Peaceful Uses of Outer Space (UNCOPUOS), which examines and monitors how well the Treaties are working, and considers the need for further legislation. This chapter provides the status in August 2023 of this regulatory material. The international space treaties are in summary:

The Outer Space Treaty of 1967 (full name: the *Treaty on Principles Governing the Activities of States in the Exploration and Use of Outer Space, including the Moon and Other Celestial Bodies*), consisting of 27 Articles, establishes, in summary, that space exploration and use should be for peaceful use and for the benefit of all mankind. It also makes clear that no country can claim ownership [of the Moon]. There is an interesting notion that astronauts are all "envoys of mankind"—not sure how that applies to space tourists—but

D. Webber, *Lunar Commerce*, https://doi.org/10.1007/978-3-031-53421-8_15

it is intended to ensure that they get assistance when needed in emergencies. The treaty makes clear that nations, not commercial entities, have to carry the responsibility of enforcing the treaty obligations. And it also places an obligation to avoid interference with others.

The Rescue Agreement of 1968 (full name: *Agreement on the Rescue of Astronauts and Return of Objects Launched into Outer Space*) establishes, in summary, that in the event of accident, or distress, or emergency situations, astronauts will be rescued and returned to base.

The Liability Agreement of 1972 (full name: *Convention on International Liability for Damage Caused by Space Objects*), in summary, establishes that states are responsible for damage caused by any of their space hardware, and must provide compensation, via an International Claims Commission.

The Registration Convention of 1975 (full name: *Convention on Registration of Objects Launched into Outer Space*) provides, in summary, for all nations placing objects into space to register some basic information regarding the objects with the UN. The data required includes name, launching state, date, orbital parameters, and purpose. Work is ongoing within such places as the volunteer Registration Working Group of ForAllMoonkind, to which the author contributed in 2018, and the GEGSLA group of the Moon Village Association (GEGSLA, 2023), to figure out if, and how, this registration requirement may need to be updated to handle the situation of commercial development of the Moon, and possible interference between different entities having permanent residence on the lunar surface.

And the aforementioned Moon Agreement of 1979 (full name: *Agreement Governing the Activities of States on the Moon and Other Celestial Bodies)* did not obtain the support of key space states. Some aspects of this failed agreement have, however, found their way into the Artemis Accords, discussed below.

It seems likely, given the overriding nature of the problems to be addressed, that a new international regulatory agency will be needed for the Moon. Such agencies exist and have been very successful in monitoring and allocating such global commons as radio frequencies and slots for satellites in the geostationary orbit (both of these examples are managed by the International Telecommunications Union (ITU) based in Geneva). The GEGSLA group, created by the Moon Village Association, as mentioned in the previous chapter, is looking into what this might entail.

What kinds of activities will need to be adjudicated and managed by such an entity? One can imagine that there will be the need for a registry of locations on the Moon where activities are taking place, maybe special environmental zones, the means of protecting lunar legacy sites, safety zones, the administration of the allocation of scarce finite resources (such as water on the Moon, or even the entire far-side), interference protection, and the provision of support and safety services (such as rescue/fire). And how do we take into account the concerns of those entities (religious, cultural, etc.) who do not want to see the Moon being used for commercial purposes?

One set of concerns is already being addressed, at least in the USA, by both some of the provisions in the Artemis Accords, and by the passing of the "One Small Step Act" of 2021 (One Small Step Act, 2021). This protects the lunar legacy sites on the Moon from future disturbance, and was the consequence of the activities of the volunteers within the "ForAllMoonkind" organization, of which the author was on the Leadership Board. The Act requires that NASA must take note of the joint NASA/Smithsonian 2011 report "NASA's Recommendations to Space-Faring Entities: How to Protect and Preserve the Historic and Scientific Value of US Government Lunar Artifacts" (and any successor reports). That 2011 report establishes suggested protective zones around artifacts to ensure that future landers or tourists do not create disturbances to their pristine state. These suggested zones and their dimensions were estimated based on the knowledge at that time about the likely energy levels of regolith particles disturbed during descent to the surface, and during hopper activities, or high-speed rover traverses near the heritage artifacts. The recommended distances for landers were 2 km for Apollo sites, and for rovers there were exclusion zones reduced to 75 meters, and also speed limits. It is of some interest to note that this activity began as a consequence of the activities of the Google Lunar XPRIZE, which we mentioned earlier. The author was Vice Chair of the international team of volunteer judges for that competition, and noted the risks of one of the potential prizes, which required the potential GLXP Team to get close to a legacy site and take a HiDef photo to earn an incremental $4 M "Heritage Prize." So, after discussions in 2013 with those behind the NASA/NASM report of 2011 (NASA/NASM, 2011), the judges inserted a requirement into the competition rules to ensure that no damage would be done in pursuit of that particular prize. A TED-talk was given to the assembled GLXP Teams in Budapest at that time for an annual conference, to ensure they understood the new rules. At that time, it was estimated that there were about 70 potential lunar surface legacy sites, and an associated 1000 artifacts, which needed to be documented.

The information needing to be recorded would be important for a number of reasons in the future, including for the protection and alert of future lunar surface tourists. After the GLXP competition ended (March 2018), the work of protecting the legacy sites was taken up by the newly formed ForAllMoonkind group who, in June 17, 2019, addressed the assembled UNCOPUOS delegates about the importance and urgency of this work, due to the imminent return to the Moon of multiple missions, including the Artemis program. We believed that this was generally well received—but that might just have been a consequence of the excellent free lunch provided in the venue. The work of ForAllMoonkind, at the UN's COPUOS, continues toward a full international enactment of such provisions, embracing artifacts delivered to the lunar surface from all states or commercial entities, which are deemed to be of overriding historical significance (ForAllMoonkind, 2023). Draft data categories for the lunar registry which were determined by the Registry Group of ForAllMoonkind, go well beyond the registration needs of the 1975 Registration Convention, and for instance include the need to make note of hazmat and nuclear elements associated with early lunar landing spacecraft. It will surely take many years for such matters to receive international agreement. Meanwhile, the plans of the several nations contemplating returning to the Moon press on.

One promising idea, floated by the MVA for protection of the Moon's pristine state, is the notion of, at least initially, limiting lunar occupation and commercial mining and manufacturing activities to a special internationally *agreed experimental zone*, thus leaving the majority of the Moon's surface in its pristine state. Some of my colleagues among the GEGSLA Observers group refer to this as the "lunar sandbox" idea. At present, this idea has not been tested within the UNCOPUOS framework, although it is being discussed within the aforementioned GEGSLA group, as a precursor to having it discussed in the full Vienna UNCOPUOS forum. It might be hoped that this, together with making clear that there would be no visual impact to the appearance of the Moon from Earth with the naked eye as a result of any proposed mining activities, may help allay at least some of the concerns of those not wishing to see lunar commerce develop.

There are, furthermore, some legacy concepts which continue on from provisions in existing Treaties. Probably the most problematic is the administration of the concept of "benefit sharing." The essence of the issue is the balancing of the mutually conflicting needs between, on the one hand, the need for commercial entities to make profits in a competitive field, and therefore maintain trade secrets, with, on the other, the requirements of treaties to share information, and indeed share the benefits with all treaty signees. One means to squaring this circle has been proposed within GEGSLA, and would involve merely offering all states an opportunity to "share the benefits" by *becoming part of the value chain* of constituent businesses operating on the Moon. If this were accepted, then the value chain information in this book could be a helpful resource.

In parallel with all of this international lunar regulatory activity, there will need to be the regulatory agencies at a national level, which will oversee how the global laws are to be interpreted domestically. Each country will designate its own preferred agency, or agencies, to oversee this work. In the USA, for example, which has over 400 agencies and subagencies, currently the Federal Aviation Administration (FAA) carries some of these responsibilities. Others are conducted by the Federal Communications Commission (FCC) and the International Trade Administration (ITA). Some now even falls within the remit of the newly constituted US Space Force. Because international agreement by total global consensus takes so much time, and because the USA has begun a process of returning to the Moon in the next few years, the *Artemis Accords* (Artemis Accords, 2023) were introduced as bilateral documents between NASA (USA) and corresponding national space agencies from other countries. At present, 28 countries have signed the Accords, which are summarized in Appendix C. Although not a true international treaty, they do provide some intercountry agreements for an interim pathway forward for commercialization of the Moon, and do specifically contain a provision for the protection of lunar legacy sites. Perhaps, the most significant national-level regulatory activities which are relevant to the contents of this book, beyond those put in place by the USA, originate in Luxembourg, where they have been making significant strides to make mining of astronautical bodies legitimate, with a particular focus on asteroids. There is now some mirror-language in the US domestic space law canon (Space Act of 2015).

The detailed and interconnected material about future lunar commerce markets, as developed in the Lunar Commerce Portfolio, and as presented in this book, will form a good background status source, updated regularly, of the direction of the lunar economy and its potential value. And, as such, will be beneficial to the administrators within the affected international and national agencies.

References

Artemis Accords. (2023). NASA.gov/specials/Artemis-accords/index.html

ForAllMoonkind. (2023). https://Forallmoonkind.org

GEGSLA. (2023). *Global expert group on sustainable lunar activities.* https://moonvillageassociation.org/global-expert-group-on-sustainable-lunar-activities-GEGSLA/

International Space Law. (2017). *International space law: United Nations instruments.* UN Office of Outer Space Affairs.

NASA/NASM. (2011). NASA's recommendations to space-faring entities: How to protect the historic and scientific value of US Government Lunar Artifacts.

One Small Step Act. (2021). https://spacepolicyonline.com/news/president-signs-law-protecting-lunar-heritage-sites/

Space Act of 2015. Spurring private aerospace competitiveness and entrepreneurship. Wikipedia – Commercial Space Launch Competitiveness Act of 2015.

16

Conclusions

We set out at the outset to attempt to understand the likely scale and timing of the initiatives involved in creating a true lunar economy. We wanted to understand the subelements of lunar commerce. This is because the stated aims of having a permanent presence on the Moon would not be possible without a lunar commerce component. And sustainability is indeed a key element in the space policy area today. At a more basic level, we have been trying to answer those questions posed by the man/woman-in-the-street. Questions about why are we doing any of this at all, and what are the benefits. We have tried to understand what we have been doing in space, and why. Why the early astronauts took those risks. What we learned from the science. What has improved over the last 50 years, and whether we can really do business on the Moon. And, moreover, we have tried, via the careful work of my colleagues who built the Lunar Commerce Portfolio, to provide a reality check of where we stand today. Did we succeed?

We did come up with some clarifications of rationale, which our typical observer might find satisfactory. Basically, we do it because we must. The only variable is the timescale. We have not been able to pin down a timescale which justifies any level of investment in government space programs greater than that which exists today. We presented an argument about a "window of opportunity" which would suggest that we ought to proceed with haste. But the rationale is in all honesty problematic. We certainly know that we must do the work leading to some eventual human settlement off-Earth. And we established that to do that will involve the creation of a lunar economy (because the Moon is the obvious local destination for trying out the necessary skills). And we established a framework of what would be needed to create that mutually-supporting lunar commerce infrastructure. But it could well be multiple

D. Webber, *Lunar Commerce*, https://doi.org/10.1007/978-3-031-53421-8_16

decades into the future before it will happen. It is perhaps a major finding of this work that it will *not* be possible to establish a true lunar economy in the remaining years of this decade (i.e., in the Early Phase), and so the governmental rhetoric needs to be adjusted to take this into account. But yes, I think we have been able to create the elements of a lunar marketplace. Did we convince you? It needs more work, that's for sure.

We also discussed the issues summarized by the question of should we be allowed to do it at all. We presented the international and national perspectives on that. Some people take a romantic view of the Moon, and say it should be left as-is. This is "the Moon is for lovers" hypothesis. It seems clear that some form of international oversight will be needed to monitor the various new regulations that will be needed for this new domain of mankind's endeavors. There could well be a regulatory compromise that, while protecting the essential surface features unchanged, it might be possible, e.g., to set aside an *experimental area*, where the necessary technology, experimentation, development, and training can take place. This book, and in particular the Lunar Commerce Portfolio which it introduces, provides an updatable snapshot of the potential lunar activities that may emerge as the constituent elements of a lunar economy. It provides some kind of basis for space agencies to use when contemplating joint government/commercial activities in the future development and exploration of the Moon.

What about the numbers? Isn't that what this was all about? What of Francis Bacon's dictum? We certainly started with doubts—did we indeed end with certainties? I don't think so. Probably not enough for most VC investors. Not yet. We used the findings of the Lunar Commerce Portfolio to come up with revenue estimates for the Mature Phase of lunar commerce which are of the order of $31 B per year. But that was not really what this was about. Because we also found out the high level of uncertainty in these findings, due to our current inadequate knowledge of a number of key parameters. And so, we identified (with certainty!) the particular areas where further research and experimental test results are needed to reduce the risk levels. And we also identified the policy areas needing some thought, and which, while remaining unresolved, contribute greatly to the uncertainties, and therefore to the risks for potential investors. Work will continue using the data and model of the Lunar Commerce Portfolio Version 1 to advance to future versions which have no missing data elements, and where all the data used carries less inherent uncertainties. Bocconi University's Business School is helping get this done. So, it was really about developing an understanding, and setting up a structure, to make a firm basis for future developments. And I think we did that.

This has been a primer. There is vastly more information in the LCP itself, and in its associated model. At the time of writing about 5000 downloads of that free material have taken place, and so a new structure has become accepted for studying lunar commerce. This book provides you, among other things, with the key to unlocking that material, so that you can contribute to that better understanding. New work on the same theme has begun as this book was being completed in August 2023—the US DARPA agency has commissioned it under the heading of LUNA-10, with findings due for issue in June 2024. This funded DARPA contract will be conducted under an Other Transaction Award category of up to $1,000,000. Which should be enough to provide improvements on the wholly-volunteer-derived material we have been describing in this book. This was the start, and we look forward to many more endeavors that improve on these initial findings.

So, you are now equipped to put on your lunar boots and be part of this great new future of possibilities opening up on the Moon. There is no reason on Earth why not.

Appendices

Appendix A: Lunar Tourism Assumptions (Used for Lunar Commerce Portfolio Development)

You don't need to read this—that's why it's in the Appendix—but just be aware that it *does* drive almost all the forecasts. Space tourism and private access to space in a terrestrial setting was an important driver of the need for reusability, resulting in increased safety and lowered costs, for space launch in general. If we are going to establish a market for lunar business, then it is an important step to establish a market for lunar tourism, which would be, it is assumed, a likely driver of much lunar activity. Is there indeed such a market? Are there enough wealthy people who would want to go and do it? Can we figure out a reasonable range of possibilities, given what we know from previous surveys, and from current knowledge of wealth statistics?

The analysis below, complete with its own references, was developed to provide a set of common assumptions for use across all market sectors during the work of creating the Lunar Commerce Portfolio, Version 1 issued in November 2022. It will be seen that, although thoroughly researched, and supported by careful analysis, nevertheless there is a distinctive lack of sound, statistically valid, and recent market survey data to underpin the work, and so this is an area of research which needs to be undertaken for future updates of the Lunar Commerce Portfolio. The material is provided in this Appendix as a marker of the Version 1 assumptions, and also as a record of the forecast methodology (based largely on the approach which was used for the

© The Editor(s) (if applicable) and The Author(s), under exclusive license to Springer Nature Switzerland AG 2024
D. Webber, *Lunar Commerce*, https://doi.org/10.1007/978-3-031-53421-8

foundational Futron/Zogby space tourism demand estimation work of 2002—(Webber et al., 2002)), which methodology can be reused once better market research findings are available. For the time being, *it represents a first attempt at a lunar tourism demand forecast, based on sound principles, economics and some admittedly rather old market data.*

The following text presents the original analysis, and its associated reference sources, that was presented to, and was used by, the volunteer analysts of the MVA Working Group on Lunar Commerce and Economics, as they prepared their respective market sectors for the Lunar Commerce Portfolio. In this way, it was ensured that all sectors were using a coherent set of driving assumptions as they developed their respective demand estimates in November 2022. It has not been updated to reflect any changes that might have occurred since then.

1. Introduction

Since it may have a significant impact on demand for various commercial services in a future lunar economy, we need to have an estimate of the likely scale of the potential steady-state future lunar space tourism business, as applicable for the Mature Phase (i.e., the assumed steady-state period beyond 2040 at the earliest). A fairly comprehensive search of the literature discovers that there has been not much in the public domain to address this question. There has been a series of market surveys carried out to assess the revenue potential of terrestrial space tourism—both of the orbital and suborbital kind. But not so much regarding lunar space tourism opportunities and demand. For the Early Phase—the period up to 2030—we can assume zero lunar surface space tourism opportunities, and only a few isolated trips into lunar orbit (with those tourists being serviced entirely with their needs by provisions brought from Earth).

Space tourism in general has proven to be of great interest, and both suborbital and Earth-orbital missions have been oversubscribed. Recent suborbital space tourism flights (July 2021), by both Blue Origin and Virgin Galactic, are reported to have been priced at over $200,000 or more per passenger, while the Earth-orbital missions, using Russian Soyuz spacecraft arranged by Space Adventures between 2001 and 2009, cost between $20 M and $70 M each, at the time. A recent Dragon flight to the ISS, arranged by Axiom Space, had ticket prices of $55 M per seat. A recent report from investment bank UBS (Globetrender blog, 2019) puts a value of $3 B a year on currently estimated space tourism markets. The Industry Arc analytics company (Industry ARC, 2021) anticipates a compound growth rate in space tourism revenues of

12.4% during 2020–2025, and they record that more than 500 people signed up with Virgin Galactic at a price of $250,000 per ticket for rides on the Virgin Galactic suborbital spaceship from Spaceport America in New Mexico.

Lunar space tourism is therefore merely the latest stage in the development of the space tourism business, which is valuable in its own right as the latest stage in the development of the terrestrial based tourism sector (which is a major economic engine in the terrestrial economy), and furthermore is destined to possibly become a major driver of the commercial development of the Moon. In this document, we consider in sequence two distinct types of lunar space tourism: lunar orbit space tourism (which, as we shall see, already has some customers signed up), and lunar surface space tourism, which is assumed to come on the scene later, as the associated technologies develop during the Mature Phase. First, we look at lunar orbit space tourism.

2. Lunar Orbit Space Tourism

Apollo 8 was the first (in December 1968) crewed mission to make the journey to the Moon, go into lunar orbit, and return to Earth (Webber, 2020). Since then, there have been a few plans put forward to send space tourists into lunar orbit. In 2007, Space Adventures offered the DSE-Alpha mission using the Russian Soyuz governmental spacecraft, linked with a specially built mission module, for two tourists accompanied by a trained cosmonaut, at a price of $150 M per tourist (Wikipedia, 2021). Subsequently, in 2017, SpaceX offered a mission using a private Dragon capsule launched on a Falcon Heavy launcher, to also be offered at $150 M per person. However, before this SpaceX mission could become established, SpaceX changed the parameters of the offering. SpaceX President Elon Musk announced that the Falcon Heavy would not be used for these crewed missions; it became the plan to instead conduct the lunar orbit mission using the Starship vehicle. The target date is currently still set at 2023, and a contract was signed in September 2018 with a customer for the mission. Yusaku Maezawa, a Japanese billionaire, will take with him up to six to eight of his artist friends. The ticket price for this "DearMoon" Project has not been announced (Wikipedia, 2021).

There have not yet been any statistically valid published forecasts of lunar orbital space tourism demand. Reference (Mihalic & Gartner, 2013) indicates maybe only two passengers per year at a price of $150 M each from The Space Tourist's Handbook by Anderson and Piven, published back in 2005. Some survey data on lunar space tourism aspirations was acquired as part of the Adventurers' Survey (Webber & Reifert, 2006). This information was, furthermore, included in (Webber, 2021). For the purposes of that survey (which canvassed 1000 people on the website of Incredible Adventures), the "Round the Moon" trip was defined as: "This adventure takes you away from Earth on a 3-day journey to the Moon. After going around the Moon six times (i.e., 12 hours), you will fly the 3-day return journey back to Earth. This will be an experience just like the early Apollo missions, but will not include a landing on the Moon." At a gross level of demand, where pricing does not matter, 47% of respondents "wanted to experience" the "round the Moon" mission. When asked about a "fair price" for such a mission, respondents suggested a range between $1 M and $100 M. At the low end of $1 M ticket pricing, 19% say they "would and could" go. These quantified responses were supported by a great many free-form comments among the respondents, such as "Around the Moon?—beyond my wildest dreams!", "If you are going to take the time to train—why not go for the Gold," and "I want to feel the exhilaration of seeing the Earth rise over the Moon."

It should be noted, however, that the respondents to this survey were self-selected to be adventurers, though were not necessarily wealthy individuals, and therefore, the results cannot be assumed to be statistically valid for the population at large, or for those wealthy enough to afford the ticket price. By comparison, the statistically valid Futron/Zogby millionaires survey of space tourism intentions (Webber et al., 2002) recorded "willingness" factors of 12% for suborbital flights and 10% for Earth orbit experiences at their associated price levels, when the 450 wealthy respondents had been fully informed of the risks, by identifying as "definitely likely" to the survey questionnaire. These results were designed to be valid within a margin of error of +/- 4.7% at the time. So, it would be more prudent and conservative to half these values, and therefore assume a value of about 5% for our purposes, as a combined "willingness factor" for the more expensive Lunar space tourism experience as applicable to the general millionaire population, which would take into account not just wishes, but other factors such as health limitations which could limit the options of some potential Lunar space tourists. This assumption would be in line with our stated intention to involve "no hype" in our projections. We shall need to allow for only limited lunar orbit tourism during the Early Phase, and none at all of the lunar surface variety.

3. Lunar Surface Space Tourism

No humans since the Apollo governmental astronauts have landed on the Moon, the last mission being Apollo 17 with Cernan and Schmitt, returning to Earth on December 19, 1972. However, there has been progress in terms of private access to the Moon's surface. The Google Lunar XPRIZE offered $40 M in prizes for nongovernmental teams able to conduct a mission to the Lunar surface, drive 500 meters. and send back HiDef images. Team SpaceIL, from Israel, made an attempt, launched in February 21, 2019, slowly making its way to the Moon over the succeeding weeks until April 11, 2019, when it hard-landed. The Team was awarded $1 M for the achievement by the competition organizers, even though the formal Google Lunar XPRIZE competition had by then ended, and the soft-landing was not achieved. NASA has subsequently awarded contracts to a number of former Google Lunar XPRIZE teams to conduct robotic landings under the CLPS program, as precursors for the eventual crewed governmental landings envisioned under the Artemis program. NASA has awarded a contract of $2.9 B to SpaceX to use its Starship as the landing stage of the Artemis program, with a scheduled first crewed landing now announced as "not before end 2025."

Although there were no specific quantified questions in the Adventurers' Survey (Webber & Reifert, 2006), there was an abundance of circumstantial free-form commentary about the wish to be part of a lunar surface space tourism adventure, e.g., "Would really like a surface stay on the Moon," "I'd rather land on the Moon," "Want to walk and stay on the Moon," "I want an eight or nine day Moon adventure, orbiting, landing, and walking. Maybe staying two or three days at a Moon base," "A Moon landing would be even better!", "A Lunar landing with a one-week stay is my ultimate goal, assuming that I can be successful enough to afford the trip, or that it is part of my job to welcome space tourists to a lunar outpost," "Lunar surface exploration is the thing," and "Would prefer a Lunar landing rather than just a circumlunar loop." For these reasons, a "willingness factor" of 5%, among those wealthy enough to afford the trip, is also assumed to apply to the surface lunar space tourism variant.

There has been to date just one proposed commercial lunar landing space tourism mission, proposed by the Golden Spike Company, in 2010, with a charge of $750 M per seat (Wikipedia, 2021), but that company went out of business and the website closed down in 2015.

4. Demand Calculations

Given the data as reported above, how can we assess the likely steady-state Mature Stage market demand for lunar space tourism, both for lunar orbital and lunar surface experiences? The following procedure follows the rationale which was used for the Futron/Zogby space tourism demand forecasts, and which is described in Webber et al. (2002), Mihalic and Gartner (2013) and Webber (2021). It is based on, first of all, calculating that proportion of individuals who would be wealthy enough at a given ticket price, and then, among that select group, finding by carefully conducted market survey who would be wanting the experience. It is a combination of an affordability analysis and a statistically valid market survey of intentions.

So, the approach starts with an assessment of numbers of high-net-worth individuals, with data provided from Credit Suisse Research Institute (2021), Wikipedia (2021) and Forbes Billionaires (2021). In summary, there are 2700 billionaires and 56 million millionaires in the world today. We need to have a distribution profile of different wealth ranges from $1 M net worth and upward. The number of millionaires represents globally the top 1.1% of the adult population (and in the USA, that percentage rises to 8.4%). So, we are trying to understand the breakdown of that small percentage, and it is notoriously difficult to obtain reliable wealth data for high-net-worth individuals (probably for tax-related reasons). From Credit Suisse Research Institute (2021), however, we find the following breakdown of the numbers of the 56 million wealthy individuals with net worth above $1 M:

$1–5 M   49.0 million
$5–10 M   4.5 million
$10–50 M   2.3 million
$50–100 M   0.1 million
$100–500 M   0.06 million
$500 M + 0.005 million

And to break down the numbers even further, we find from Forbes Billionaires (2021) that, breaking down the 5332 people with wealth greater than $500 M, there are 2755 billionaires (with 724 of them being in the USA).

Then, we need to estimate the likely range of price levels for both the lunar orbit and lunar lander missions. This will depend on the likely architectures—and in particular, lower prices would be the result of Starship, rather than SLS missions. We need therefore to make the calculations at both ends of the price level range. We are guided in this by comments made by Elon Musk (Musk, 2017; Musk, 2019). Thus, "similar to ISS" is about $55 M now, and so we can use that price in our range estimates below for the Mature Phase time period:

Prices for Lunar orbit mission—assume range from $55 M to $150 M per passenger.
Prices for Lunar lander mission—assume range from $290 M ($2.9 B divided by 10 passengers, ignoring at this stage the eventual $1/2 M Elon Musk

announced Mars target indicator price) to $750 M per passenger. Note that it is assumed for the purposes of this exercise that the proposed price ranges include space transportation, provision of space suits, and the accommodation and food and life support at the destination.

Then we proceed to derive demand forecasts. We know from the millionaire preference data in Webber et al. (2002) that high-net-worth individuals will typically not spend more than 10% of their net worth on an individual discretionary purchase leisure project (as was the case for the first few orbital space tourists, back in the early years of the 2000s). So, we can apply this percentage in reverse to calculate how many of the population would be able to even afford the prices of tickets for both lunar orbit and lunar surface missions, if they should want to do it.

For the lunar orbit missions, at the stated ticket prices assumed to represent 10% of net worth, this implies the wealth needed at each ticket price would be either $550 M or $1.5 B. From the wealth population statistics, this implies a potential pool of 2000 to 62,000 sufficiently wealthy people globally. If we apply the "willingness" factor of 5%, this results in a potential annual steady-state market for lunar orbit space tourism of approximately 3200 persons (at $55 M per ticket) down to 100 persons (at $150 M per ticket). The associated revenue potential range is $175 B down to $15 B. We can assume these numbers for the Mature Phase. During the Early Phase, we only assume a handful of tourists going into lunar orbit.

For the lunar surface space tourism missions, at the stated prices assumed to represent 10% of net worth, this implies the wealth needed at each ticket price would be either $2.9 B or $7.5 B.

From the wealth population statistics, we find that we are down to very small numbers as a potential pool of sufficiently wealthy candidates. The resulting range is from 200 to 2500 at the appropriate price levels. If we then apply the "willingness" factor—also assumed to be at 5%—then this results in a potential annual steady-state market for lunar surface space tourism of approximately 125 persons (at $290 M per ticket) down to 10 persons (at $750 M per ticket). The associated annual revenue potential is $35 B down to $8 B. These are the figures for the Mature Phase calculations.

We also make this assumption for the record, without at this time any data to back it up (could be a subject for future market research survey), that these two forecasts (i.e., lunar orbit and lunar surface space tourism demand) are independent of each other, and are therefore additive with respect to revenue stream contributions.

Thus, *what we have derived is based upon a series of conservative assumptions:*

(a) Assessing the likely price ranges for the offerings;
(b) Multiplying the prices by ten to arrive at the wealth level needed to even consider the experience;
(c) Finding out from wealth statistics how many people have that much wealth; and then
(d) Taking 5% of these numbers as the potential revenue generating Lunar space tourist projections.

Thus, we note that the numbers are conservative, and do not take into account any potential tourists coming from wealth categories lower than the assumed cut-offs, even the lowest level of which is a net worth of $550 M.

5.   Lunar Commerce Portfolio Market Impacts

As we noted early in the process of developing the Lunar Commerce Portfolio, and created the teams to analyze the commercial market demand potential, activities related to Lunar Space Tourism would have impacts which could potentially have implications across all the market sectors. We noted the need to record these specific impacts, while making sure that we were not double-counting the revenue opportunities. So, in this section, we proceed to track the meaning of the forecast range developed under Section 4, and allocate the impacts across the various market teams, to ensure internal consistency within the overall quantification process.

But first, we need to make some further assumptions about the way in which the forecasted space tourists are likely to arrive in the lunar environment. Specifically, how frequently should we assume the arrivals and departures, and how many folks at a time are likely to be staying in lunar orbit tourism hotels and on the lunar surface? The best data source for this is Webber et al. (2002) and Webber and Reifert (2006) (to some extent repeated in Webber (2021)), where we learn that millionaires in general are busy people, who do not spend long periods on vacation. So, we make the assumption that lunar stays, whether in lunar orbit, or on the surface, will not exceed two weeks (with another week assumed during transit). In the case of the surface hotel guests, we can make the further assumption that they will be in residence there during the period of the Lunar day, and that the hotel will be unoccupied each time during the subsequent two Earth weeks each month.

With these assumptions, we derive the following rounded numbers of space tourists at any one time occupying the lunar orbit or lunar surface tourism facilities, during the steady-state Mature Phase:

For Lunar Orbit: At the high end, the 3200 figure translates into **130 Lunar Orbit space tourists**, each staying for two weeks, then being replaced by a similar contingent. At the low end, the 100 figure translates into just **5 Lunar Orbit space tourists**, each staying for two weeks, and then being replaced by a similar contingent, and so on throughout the year.

For Lunar Surface: At the high end, the 125 figure translates into **10 Lunar Surface space tourists**, each staying for two weeks, then being replaced after a 2-week gap by another 10. At the low end, the 10 figure may be translated into **only 1 Lunar Surface space tourist**, staying for about 2 weeks and then being replaced by the next customer, again, following a 2-week gap, with this process continuing throughout the year.

Now, we know the impact assumptions of lunar space tourism in round numbers for each of the market sectors during the Mature Phase of operations. Indeed, there are impacts across the entire lunar commerce portfolio.

6.   References

6.1   Webber, D, et al, Space Tourism Market Study, Futron/Zogby, Oct 2002

6.2   Webber, D and Reifert, J, The Adventurers' Survey, Incredible Adventures, Nov 2006

6.3   Mihalic and Gartner, Eds, Tourism and Developments, Nova Books, 2013

6.4   Webber, D, Space Tourism Business - The Foundations, Curtis Press, 2021

6.5   Webber, D, MVA Webinar, Lunar Space Tourism: Getting There, Oct 2020

6.6   Webber, D, ISU SSP21 Presentation: Moon Market Development and Evolution, Jul 2021

6.7   Wikipedia, Tourism on the Moon, Accessed 24 Nov 2021

6.8   Wikipedia, Dear Moon Project, Accessed 24 Nov 2021

6.9   Globetrender blog, Space Tourism Could Generate $3 Billion a year by 2030, March 2019

6.10   Industry ARC, Space Tourism Market Forecast 2020-2025, website summary, Nov 2021

6.11   Credit Suisse Research Institute, Global Wealth Databook 2021

6.12   Wikipedia, List of Countries by the number of millionaires, Accessed 25 Nov 2021

6.13   Forbes Billionaires: The Richest People in the World, 2021

6.14   Musk, Elon, Remarks on Twitter, Feb 27, 2017, "Ticket prices to Lunar orbit similar to tickets to the ISS"

6.15   Musk, Elon, Remarks on Twitter, Feb 11, 2019, "Elon Musk expects SpaceX ticket to Mars will cost $500,000"

So, that was the formulation presented to, and used by, the analysts working on the Lunar Commerce Portfolio, during 2021 and 2022. Only two significant events have occurred since then and by August 2023 which might have influenced the original document, one being potentially positive, the other being potentially negative. On the positive side of the balance, another lunar orbit tourism trip has been contracted, in addition to the Maezawa event described in the original document. Dennis Tito (first space tourist) and his wife have made an agreement with SpaceX to be passengers on a subsequent circumlunar trip. On the negative side of the balance, Titan, a "tourist submersible" its crew and tourist passengers were lost in a dive to the Titanic, and parallels were drawn in the press to space tourism ventures.

Lunar space tourism, under our assumptions, will be a key part of our efforts to create a lunar economy. And furthermore, it is definitely real. The two photos (Figs. A1 and A2) show the signed-up crews for the first two commercial lunar orbit missions, using SpaceX as the service provider. These folks are in the vanguard as true adventurers. Both Maezawa and Tito have previously paid for, and presumably immensely enjoyed, space tourism experiences in Earth orbit. Tourists will always be pushing the boundaries.

Fig. A1   Maezawa and his assorted artist friends prepare for lunar orbit mission. (Credit: DearMoon.Earth/Spacetoday Inc)

Fig. A2 Tito and his wife at SpaceX launch site. (Credit: CBS News)

Appendix B: External Factors Affecting Forecasts

This may be a bit "heavy duty," if you are only interested in the "bottom line" numbers, but this Appendix does point out why those "bottom line" numbers are not as robust as you might otherwise think.

"The slings and arrows of outrageous fortune" have been well known, even before Shakespeare made them so memorable. In attempting to create a thriving human enclave and trading estate on a new world, such as the Moon, those planning new commercial businesses will be trying very hard to make sure that everything necessary for success will be in place. However, as we know from experience, there will be factors beyond the control of any one business venture which will act to sway the outcome, one way or the other. As we attempt to provide our forecasts for the lunar economy, the best we can do is to be aware of these external variables, and at least try to make sure that each market sector will react in unison to the emergence of these extraneous factors (although they will act in different ways—some kinds of outside impact will provide a positive influence in one market area and a negative one in another). So, we aim for *consistency*, and therefore need to have a consistent and agreed description of each of these external global factors. For our purposes within the LCP forecasting process, it was left to each market team's experts to decide in what way their market forecasts would be impacted by the changes to global common external factors. We acknowledge that there will also be factors which do not affect *all* market sectors across the board, but merely impact one sector, and for our purposes, those factors are considered within the purview of each market sector in turn. We consider here, therefore, only those factors which are extraneous, and which are *so significant that they will have an impact across the entirety of the lunar commerce markets.*

Within the Lunar Commerce Portfolio modeling, we handled the potentially complex impacts of each of these factors across the board, by considering various combinations of the 21 factors in groupings we identified as scenarios. We developed four of them (Alpha, Beta, Gamma, Delta) for the first version of the LCP, but future users can create other combinations to better approximate the anticipated geo-political environment. As shown in Fig. 6.3, Scenario Alpha was the least aggressive, and Scenario Delta was the most optimistic framework we considered.

The factors themselves were developed by consensus in a specially designated international task force of volunteer analysts within the Lunar Commerce and Economics Working Group of the Moon Village Association. The task force initially came up with the 21 factors, and the range of values at each end of the spectrum from "conservative" to "ambitious," and subsequently organized them into five main categories: Regulatory, Political, Access Architectures, Demand Factors, and Supply Factors. We describe each of them below, for the record, as they were perceived in November 2022.

Regulatory Factors

– *Degree of International and National Regulatory Controls.* The issue being addressed here is in broad brush terms whether the community—essentially the international community, but supported by national regulators—decides to encourage or discourage the return to, and commercialization of, the Moon. The main international entity is likely to be the UN COPUOS in Vienna, but other international regulators might include the ITU. At one end of the spectrum of possibilities, there may be agreement that the Moon might be used in any way as a resource for mankind. But at the other end there may turn out to be objections by some, or all, nations, or indeed within nations, which would impose severe limits on what might be allowed to take place. Elsewhere on the spectrum of possibilities, there might be a case for allowing development on the Moon, but only in a limited part of lunar geography, as a kind of "testing lab" for mankind's further long-term exploration efforts. The assumed value for modeling purposes ranges from "conservative" to "ambitious."
– *Inclusion of Peaceful Security Activities.* The reason for including this as a specific issue is that a positive answer would imply an increased number of lunar inhabitants, who in turn would require access to all the other lunar services being provided. It is assumed that this would probably be administered by an increased level of staffing in the governmental sectors on the Moon. The arrangements for security on the lunar settlement are liable to be the subject of international debate, and may be expected, therefore, to take some time to resolve. For modeling purposes, the assumed value is either "yes" or "no."
– *Some Equivalency of Land Ownership.* International law precludes the possibility of ownership of land on the lunar surface by any nation. On the other hand, in order to motivate possible future commercial endeavors, some financing institutions might look to land as a guarantee for a venture. Therefore, it is a matter of ongoing discussion whether some surrogate for land ownership on the Moon could function at the level of a commercial entity, even if not for a nation. The outcome of this debate, and precise nature of the resulting regulations, can have a significant (even profound) effect on the revenue outcome. The assumed value for modeling purposes ranges from "conservative" to "ambitious."

– *Government only option.* If it is deemed the case that only nation states may be authorized to operate on the Moon, then this would effectively put in place a major obstacle to the development of a lunar economy. The assumed value for modeling purposes is either "yes" or "no."

– *Environmental Rules.* It seems clear that, because of the mutual dependence of operators on the Moon, there will need to be some kind of common agreement about how to control the potential damage due to takeoffs and landings on the lunar surface, due to plumes of erosive lunar dust. This would take the form, presumably, of rules for landing pad surfaces and berms. By extension, similar rules would probably apply to sintered roadways. In the general category, we can also imagine the need for common standards for recycling, and operating standards for regolith mining. It is possible, for instance, that mining may only be allowed in a designated "test area" on the lunar surface. There may even be rules introduced to protect the visual appearance of the Moon from lunar orbit. Clearly, the degree to which any such rules prohibit commercial mining operations will have an impact on the revenue outcome. The assumed value for modeling purposes ranges from "restrictive" to "moderate."

– *Sharing of Scarce Resources.* This condition from the Outer Space Treaty language requires that scarce resources on the Moon must be shared. Just how this is to be done equitably, without discouraging necessary commercial developments, is the subject of ongoing debate. The beneficiaries of any such regulatory factors will be developing economies, whereas in traditional capitalist economies, the first (and therefore generally the least risk-averse) provider obtains some kind of advantage. On Earth, this is the basic concept behind the patent system. The creative risk taker expects to receive some benefit and protection for improving society by its behavior, at least long enough to recover the up-front investment costs, and ensure some profit. On the Moon, scarce resources might be assumed to encompass lunar water ice, for example. Or, one could even require that the regions of perpetual darkness at the poles would themselves be considered as scarce resources. For the purposes of modeling, we recorded for each scenario whether that required an undefined "sharing system," or whether a "first in line" approach would be assumed.

Political Factors

– *Multiple Nations Including Sino-Russian.* How will future activities in space, and particularly on the Moon, be conducted internationally? Will there be commonly designed and operated international bases, or will there be separate national bases? Will they be interconnected and therefore require common interfaces? Already, it seems that China is planning to develop its own lunar base, separate from that being proposed by the USA under the Artemis banner. It is not clear what other nations will be pursuing. How will independent commercial businesses function in either of these settings? What will be the impact on investors of either outcome? Investors require clarity above all. They need to know what they are facing. So, they will be observing developments. For the purposes of the modeling in Version 1 of the LCP, the assumed value alternatives were "collaborative" versus "adversarial" versus "neutral."

– *Zero to One or More Moon Village Bases.* The purpose of this factor is to offer the possibility of a range of supported Moon bases. The assumption is that this decision will emerge from national space agency planners, and be dependent on the level of support emerging from national taxpayers. Within each identified base area, the assumptions would likely not differ very much. However, when each extra base is introduced, this entails more roadway construction, more complex communications arrangements, and more use of lunar transportation services. For the purposes of the modeling exercise, there was an arbitrary selection of up to five bases.

– *Government Space Exploration Budgets.* The purpose of this factor is to capture any change to the baseline governmental expenditures that are currently taking place now, and which are forecasted through the remainder of the Early Phase. The question is whether they will continue at much the same base level moving into the Mature Phase, or will they change either up or down? In the modeling exercise, we were just considering whether there would be any big changes to governmental space expenditures as a share of national budgets, versus simply maintaining the same general proportions as today. For the modeling exercise, the assumed value for this factor was either "conservative" or "ambitious."

Access Architectures

– *Cost and Ease of Access.* The main focus of this factor is to explore the implications of whether the dominant means of access to the Moon will be government or private sector vehicles. Clearly, there will be a massive impact on the cost per Kg depending on this. It will affect not just the cost but also the flight frequency, payload mass and number of seats for crews and tourists. For the purposes of the modeling approach, values were assumed for "Starship" versus "SLS/Artemis," versus a "combo" of both approaches. In terms of numerical values assumed, we assumed in all cases for the Early Phase, there would be 22 annual arrivals and departures to the lunar vicinity, and for the Mature Phase, the values would be 22 for scenario Alpha, increasing to 66 (plus return payloads) for scenario Delta.

- *Adequacy of Power Sources.* On the majority of the Moon, its slow rotation rate means that for two Earth-weeks at a time, the surface will be in complete darkness, and very cold. Near the poles, there are a few special exceptions to this rule. Human survival on the Moon must include survival of the long lunar night, which implies both the need for energy storage devices and the use of nonsolar energy sources, which would include nuclear generators. Clearly, use of nuclear sources has as a corollary the need for a whole range of different behavior patterns among the residents on the Moon. Therefore, how this power is derived will have a significant impact on the costs and operations on the Moon's surface. For the purposes of the model, the main distinction of values was characterized as either "solar power only" or "solar and nuclear power sources."

- *Lunar Location Options.* This is a major consequential matter, with enormous implications for business on the Moon. Most of the Apollo landings took place relatively close to the lunar equator. The current phase of Artemis plans involve landing at the poles. What will be the ongoing development? It could well be that for instance some important resources will not be available near the poles. And lunar surface tourists might be expected to want to visit historic lander legacy sites. So, are we to assume landing only at the poles, or will there be wider deployments? There will be important differences with regard to power provision, and also with regard to mass differences for deliveries to the lunar surface. For the modeling activity, the alternative values were described as "South pole only" versus "Wider deployments." In the case of the "wider deployments," we assumed the numerical value of four independent bases.

- *Human Health Factors.* This factor has been included because at present we do not know how well humans will adapt to long periods of operation under 1/6 g conditions. We need to be concerned about resilience and ability to continue to perform operations, and to sleep well. Ultimately, we shall also need to know of any impact on reproduction. For this First Version of the LCP, this factor was included simply as a place keeper, for possible future modification, and it was assumed in all four scenarios that long-term residence on the Moon would be possible. Quite clearly, there could be dramatic impacts on lunar commerce if it turns out that there are overriding problems with that assumption.

- *Connectivity and Data.* This factor is intended to reflect the issue of how data channels will be handled on the Moon. Will policy be developed putting government in the forefront, or will commercial initiatives be allowed to flourish and become established? Will there be a private—only setup, or government, or government plus private? And what about timing? Probably important to have an initial government-built infrastructure. For the Version 1 modeling, we considered only two values—"private only" or "government plus private."
- *Private Sector Financing.* For Version 1 of the LCP, this factor was included merely as a placeholder, with the values offered simply as "limited deployment" versus "widely deployed." Further work on this factor TBD.

Demand Factors

- *Scale of Lunar Space Tourism.* As noted elsewhere in this account, it transpires that the assumptions we have adopted regarding the likely demand for lunar tourism, both on the surface and in lunar orbit, have a considerable impact on the lunar economy in general. The analysis and development of the assumptions used, with range of uncertainty, have been recorded in another Appendix. For the purposes of the modeling, we allocated values ranging from "conservative" to "ambitious." For the conservative assumption, we assumed 10 tourists a year to the surface, staying two weeks (i.e., 1 at a time, each month) paying $750 M plus 4 government support astronauts. For the ambitious end of the range, we assumed 125 tourists a year going to the lunar surface (i.e., 10 at a time, each month, staying for 2 weeks) paying $290 M plus 10 government support astronauts.
- *Degree of Lunar Orbit Activities.* This factor also records an important assumption about demand for services. To what degree will the proposed Lunar Gateway station be used as a destination for lunar tourists? Will there be enough demand so that additional stations will need to be built either attached to the Gateway, or being quite independent? Will occupants of these lunar orbiting stations or hotels be provided with food supplies direct from Earth, or will they be sourced from the Moon base as a commercial business transaction? These assumptions affect the estimates of permanent and visiting populations. For the modeling work, we used the alternative values of "conservative" or "ambitious." In numerical terms, we assumed for the conservative case 100 lunar orbit tourists/year (i.e., 5 at a time, staying two weeks) paying $150 M, plus 4 governmental support astronauts. In the ambitious case, the numerical values assumed were 3200

lunar orbit tourists/year (i.e., 130 at a time, staying two weeks) paying $55 M, plus 10 governmental support astronauts.

- *Lunar Mining for Terrestrial Needs.* A baseload of activity is assumed for the purposes of sustaining the viability of those living and working on the Moon. This would involve the mining of sufficient regolith to provide enough water and oxygen for survival, and even to provide the rocket fuel to be used for ongoing or returning visitors. But will there develop enough commercial demand to justify the costs of transporting materials from the Moon to the Earth? This is a complex question which embraces the definitions of scarcity on Earth, and strategic geography of alternative terrestrial sources. In this category, we would mainly be addressing terrestrial needs for Platinum Group Metals, rare Earths and maybe He3. For the purposes of this Version 1 of the LCP, we used the values "conservative" versus "ambitious." In the conservative case, we assumed zero as the numerical value, with 17 tons a year of PGMs assumed for the ambitious case.

- *Terrestrial Markets for Lunar Manufacturing.* The intention behind this category is to reflect the currently unknown demand for products to be created on the Moon and transported back to Earth. Clearly, this is in general a very expensive thing to do, and it could only make economic sense if there are some products that can be made on the Moon which far exceed certain quality standards, or other efficiencies, that are not possible on Earth. And this could only be the case if there are some significant inherent advantages to manufacturing as a consequence of the near-vacuum environment and 1/6 g gravitational field. It is also possible in principle that there might be some high value components, or luxury/premium consumer products, which could close a business case taking into account the supply chain economics. For our modeling purposes, we awarded the values of either "conservative" or "ambitious" to these demand factor outcomes between different scenarios.

Supply Factors

- *Exploitability of Lunar Ice.* We now consider the unknown extraneous factors related to the development of resources on the Moon. We are trying to reflect here the degree of difficulty which emerges as we begin trying to find, extract, and process lunar water ice, currently assumed to be available in quantity in the polar cold traps. Clearly, this is a supply factor which will affect everybody operating on the Moon, and the likely price levels of the resulting resources. In our comparison of scenarios, we tried to reflect a range of levels of difficulty in extraction which might emerge in the Mature Period. If indeed we are operating at the most difficult end of the spectrum, it would result in a delay in the entire timing of the Mature Phase itself. When would there be sufficient lunar ice extracted to build a prototype plant scalable for both life support and predicted propellant demand? In Version 1 of the LCP, we allocated the values as either "conservative" or "ambitious." And we assumed the numerical value of 5240 tons/year as the conservative assumption, and 22,620 tons/year for the ambitious assumption.

– *ISRU Difficulties.* This is an extension of the above supply factor consideration, intended to capture the current unknowns regarding the ease or otherwise of accessing mineral resources on the Moon. How easy will it prove to find, extract and process lunar minerals for use on the Moon, or possibly for the terrestrial needs? We also tried to capture here the need to solve problems of cryostorage, particularly for possible rocket propellant markets. We simply, for the purposes of defining our scenarios in this first use of the LCP, allocated the value of either "substantial difficulties" versus "few difficulties."

To find out how the ends of the spectrum "conservative" versus "ambitious" were treated in LCP Version 1, you are referred to the detailed assumptions contained within the EXCEL spreadsheets of the model in the publicly available resource.

In retrospect, the material described in this Appendix served its purpose. For future versions of the LCP, there could be at least three different kinds of change offered. First of all, there could be additional *factors* introduced—possibly as a result of a more systematic and knowledge-based assessment. Then there could be perhaps a better assessment of the *numerical consequences* of moving from the "conservative" to the "ambitious" ends of the spectrum, for each factor considered. And then, of course, future users of the LCP might use different *combinations* of the External Factors to represent new "worldviews," creating new and different scenarios to assess the likely consequences of alternative regulatory activities.

However, we might reflect that the main value of the analysis represented here is not so much in determining the numerical consequences, but is rather in focusing the mind on the issues that remain which are in need of resolution.

Appendix C: Summary of the Artemis Accords (Status Recorded as at August 2023)

Normally, regulations don't make for a great read, but these have been condensed, so that you get the essential points.

This Appendix is included because the new era of exploration of the Moon, and its potential commercialization, will operate within an international legal framework which to a large degree will be circumscribed by this nonbinding agreement. It will require certain restrictions in behavior, and at the same time, open up opportunities for future conduct, including potential commercial activity on the Moon. The Accords represent a common set of basic principles to govern the exploration of the Moon (and, incidentally, beyond, but we focus here on the Moon in particular). To some degree, they simply restate matters which have already been codified within the Outer Space Treaty of 1967 and the Registration Convention. But they do go further—in particular addressing the issue of resource extraction, and recognizing the need to protect cultural heritage sites on the Moon. They were established in 2020, and since then have been accepted by a succession of signatories on a bilateral basis. It is a prerequisite for any participation in NASA's Artemis lunar program to sign the Accords. In the text of the Accords themselves, focusing on lunar activities, it is stated that "Adherence to a practical set of principles, guidelines, and best practices in carrying out activities [on the Moon] is intended to increase the safety of operations, reduce uncertainty, and promote the sustainable and beneficial use of [the Moon] for all humankind."

Background

The genesis of the Accords was the decision to create the Artemis Program to return astronauts to the Moon. And the realization that to do it, unlike the case with the Apollo program in the sixties, will require international involvement, and commercial support elements. There was a realization that normal UN-led international agreements and treaties may take even decades to achieve, and even then, may not be possible at all if all nations cannot agree. Meanwhile, activities were moving ahead within NASA, and international partners were becoming involved, and something was needed to codify a framework for exploring and even mining the Moon. The decision was therefore made to move ahead on a bilateral basis. The Accords were therefore developed, and initially signed in October 2020, by the directors of NASA and a first batch of national space agencies. Other nations have subsequently added their signatures (see section "Signatories").

Content

The Artemis Accords consists of a document listing the following ten main broad aspects of activity on and around the Moon, around which it is hoped general international agreement can be achieved. The content is summarized below.

- *Peaceful Purposes.* Self-evident
- *Transparency.* Agreement to share policies and plans, and scientific information obtained [from the Moon activities] on a good-faith basis. (NB. This might cause some difficulties with regard to commercialization and commercially sensitive exploration findings obtained by investment in a competitive environment.)
- *Interoperability.* This makes sense in an international environment on the Moon. Specifically mentioned are interoperability standards for fuel storage and delivery, landing structures, communications systems, and power systems. (N.B. this only enhances the possibilities for commercial development of the Moon. Maybe spacesuit standards, at least regarding fittings for access to oxygen supplies, and water, should also be considered).
- *Emergency Assistance.* This is also self-evident, and is merely a restatement of obligations under the existing Rescue and Return Agreement.
- *Registration of Objects.* This, as written, is merely a restatement of obligations to the existing Registration Convention—i.e., to register a few basics about the rocket and spacecraft being used for a mission (N.B. There could emerge, however, a need—being explored within the MVA and GEGSLA—to further extend these requirements to register some minimal dataset representing activities taking place on the Moon—in order to prevent mutual interference between adjacent commercial operators).
- *Release of Scientific Data.* The same comments apply as were mentioned above regarding possibility of commercial sensitivity of data. Maybe some time period may be part of a future regulation to take this wrinkle into account. Commercial operators might be allowed some time in order to reap the first rewards of their investment, before having to declare the information publicly. With these points in mind, the exact language in the Accords includes "as appropriate, in a timely manner." And also specifically exempts private sector operators from the requirement, unless they are conducting operations on behalf of their national space agencies.

- *Preservation of Outer Space Heritage.* This is a new element in international space agreements, and emerged following activities during the Google Lunar XPRIZE, and as a result of subsequent activities of the ForAllMoonkind NGO (including a June 17, 2019, presentation to the United Nations' COPUOS in Vienna). The specific description in the Accords refers to the need to "preserve outer space heritage …including historically significant human or robotic landing sites, artifacts, spacecraft and other evidence of activity." (N.B. In practical terms, this will involve the need to keep an appropriate distance from designated legacy sites when landing on, moving by, or operating near them.)
- *Space Resources.* This is probably the most significant part of the language of the Accords, with regard to the activities recorded in this book, viz., regarding commercializing activities on the Moon to create a lunar economy. Focusing on the specifically lunar aspects to the language in the Accords, it is made clear that extracting lunar resources does not contravene the article about national appropriation in the Outer Space Treaty. (N.B. This is where the signers of the Artemis Accords are agreeing that it is OK for private entities to extract lunar resources—because it is not about a nation "owning" the Moon, or any part of it.)
- *Deconfliction of Space Activities.* This is the longest, most detailed section of, the Artemis Accords, having eleven separate subsections to address the issues. The main idea is to continue the requirements of the Outer Space Treaty with regard to potential harmful interference, adding in a need for notification of the location and nature of activities, and respecting a need by signatories for "due regard" of what others are doing. This section also introduces the concept of "safety zones" on the lunar surface, and the need to coordinate activities within such zones. The zones themselves can move in geography and time coordinates, and would require prior notification of such changes. (N.B. For our purposes, we can imagine conducting our lunar commerce activities while following these rules and thereby avoiding conflict on the lunar surface. It will also require the setting up of some regulatory personnel both on Earth and on the lunar surface.)
- *Orbital Debris.* We have, it has to be admitted, had only limited success in mitigating orbital debris problems in Earth orbit. But hope springs eternal, and maybe we can make a better job of the region around the Moon. This element of the Artemis Accords refers to "passivation and disposal of spacecraft at the end of their missions," without stating how. And goes on to commit to limit generating long-lived harmful debris, by taking appropriate measures such as selection of safe flight profiles. (N.B. The impact for our purposes of such provisions would be to limit some flight profile options,

and maybe reduce the effective operating lifetime of lunar orbiting facilities because of the need to use fuel—which would otherwise have been revenue-generating—to manage the end-of-life maneuvers of lunar orbit spacecraft, in whatever way is deemed appropriate).

Signatories

At the time of completing the text of this book for publishing (August 2023), there had been 28 countries and 1 territory who had signed onto the Accords:

- Argentina
- Australia
- Bahrain
- Brazil
- Canada
- Columbia
- Czech Republic
- Ecuador
- France
- India
- Israel
- Italy
- Japan
- Luxembourg
- Mexico
- New Zealand
- Nigeria
- Poland
- South Korea
- Romania
- Rwanda
- Singapore
- Spain
- Saudi Arabia
- Ukraine
- United Arab Emirates
- United Kingdom
- United States
- Isle of Man

Index